Patch Clamp Technique - Current Methods and Future Prospects

Edited by Robson Xavier Faria

Published in London, United Kingdom

Patch Clamp Technique - Current Methods and Future Prospects
http://dx.doi.org/10.5772/intechopen.1003328
Edited by Robson Xavier Faria

Contributors
Amina Asghar Chaudhry, Andrzej Teisseyre, Anna Palko-Labuz, Guilherme Pegas Teixeira, Juan Ignacio Felice, Kamila Środa-Pomianek, Leandro Rocha, Melisa Moncada, Nicolas Enrique, Pedro Martin, Robson Xavier Faria, Shazia Perveen, Silvia Santillo, Sumaira Kanwal, Toru Yazawa, Veronica Milesi

First published in London, United Kingdom, 2024 by IntechOpen
IntechOpen is the global imprint of INTECHOPEN LIMITED, registered in England and Wales, registration number: 11086078, 167-169 Great Portland Street, London, W1W 5PF, United Kingdom

For EU product safety concerns: IN TECH d.o.o., Prolaz Marije Krucifikse Kozulić 3, 51000 Rijeka, Croatia, info@intechopen.com or visit our website at intechopen.com.

British Library Cataloguing-in-Publication Data
A catalogue record for this book is available from the British Library

Patch Clamp Technique - Current Methods and Future Prospects
Edited by Robson Xavier Faria
p. cm.

This title is part of the Biochemistry Book Series, Volume 58
Topic: Cell and Molecular Biology
Series Editor: Andrei Surguchov
Topic Editor: Mary C. Maj
Associate Topic Editor: Felicia Ikolo

Print ISBN 978-1-83769-754-0
Online ISBN 978-1-83769-753-3
eBook (PDF) ISBN 978-1-83769-755-7
ISSN 2632-0983

If disposing of this product, please recycle the paper responsibly.

IntechOpen Book Series
Biochemistry
Volume 58

Aims and Scope of the Series

Biochemistry, the study of chemical transformations occurring within living organisms, impacts all of the life sciences, from molecular crystallography and genetics, to ecology, medicine and population biology. Biochemistry studies macromolecules - proteins, nucleic acids, carbohydrates and lipids –their building blocks, structures, functions and interactions. Much of biochemistry is devoted to enzymes, proteins that catalyze chemical reactions, enzyme structures, mechanisms of action and their roles within cells. Biochemistry also studies small signaling molecules, coenzymes, inhibitors, vitamins and hormones, which play roles in the life process. Biochemical experimentation, besides coopting the methods of classical chemistry, e.g., chromatography, adopted new techniques, e.g., X-ray diffraction, electron microscopy, NMR, radioisotopes, and developed sophisticated microbial genetic tools, e.g., auxotroph mutants and their revertants, fermentation, etc. More recently, biochemistry embraced the 'big data' omics systems. Initial biochemical studies have been exclusively analytic: dissecting, purifying and examining individual components of a biological system; in exemplary words of Efraim Racker, (1913 –1991) "Don't waste clean thinking on dirty enzymes." Today, however, biochemistry is becoming more agglomerative and comprehensive, setting out to integrate and describe fully a particular biological system. The 'big data' metabolomics can define the complement of small molecules, e.g., in a soil or biofilm sample; proteomics can distinguish all the proteins comprising e.g., serum; metagenomics can identify all the genes in a complex environment e.g., the bovine rumen.
This Biochemistry Series will address both the current research on biomolecules, and the emerging trends with great promise.

Meet the Series Editor

Andrei Surguchov, Ph.D., joined Baylor College of Medicine, Houston, TX, as a faculty member in 1992, where he studied the mechanisms of the genetic control of lipid metabolism. At the University of Utah, his research interests focused on cloning new genes encoding retinal proteins. He studied molecular and cellular mechanisms of neurodegenerative diseases and retinal degeneration at Washington University, St. Louis. Currently, his research focuses on the structure-function relationship of proteins involved in neurodegeneration and ocular diseases. Andrei Surguchov is an Editor-in-Chief at Biochemistry Research International and Associate Editor in several biomedical journals.

Meet the Volume Editor

Robson Faria received his undergraduate degree in Biological Sciences from the Federal University of Rio de Janeiro, in 2001, and an MSc and DSc from the Oswaldo Cruz Foundation's Cellular and Molecular Biology program, in 2003 and 2006, respectively. He has been a public health researcher at the Oswaldo Cruz Foundation, in Brazil, since 2007. During the past 20 years, Dr. Robson has been working with purinergic receptors, especially P2X7R, and the action of synthetic molecules and derivatives of natural products to discover new prototypes capable of treating neglected diseases.

Contents

Preface

Electrophysiological properties are closely related to the physical, biological, and chemical characteristics of cells. The patch-clamp technique is used to investigate cellular physiological functions by measuring electrical currents in individual cells or ion channels. This book summarizes classical and recent advancements in the patch-clamp technique, particularly in exploring the chemical and biological mechanisms of ion channels and cellular functions.

In addition, it briefly highlights the association of new techniques and instruments with patch-clamp methods and their application in cell electrophysiology studies. The text provides an overview of patch-clamp configurations and their adaptability to various conditions, simulating real conditions related to ion channels and membrane electrical properties.

I thank all the contributing authors for their tireless and essential collaboration in this project.

Robson Xavier Faria
Ph.D. Laboratory of Evaluation and Promotion of Ambiental Health,
Pavilion Ozório de Almeida,
Oswaldo Cruz Institute,
Oswaldo Cruz Foundation,
Rio de Janeiro, Brazil

Chapter 1

Introductory Chapter: Patch Clamp Technique – Current Methods and Future Prospects

Robson Xavier Faria

1. Introduction

Many important biological phenomena are accompanied by electrical and cellular manifestations. At rest, living cells have electrical potential differences of several tens of millivolts across the plasma membrane, with medium-negative intracellular fluid in relation to extracellular fluid (ECF). The genesis of this membrane potential is associated with ion transport mechanisms, which create an intracellular ionic medium with a composition distinct from that of the extracellular ionic environment. Therefore, diffusion processes (diffusion potentials) and active transport (pump potentials) electrogenic processes) represent the basic mechanisms responsible for the polarization of the plasma membrane. The diffusion of ions through concentration gradients is the most important electrical manifestation in biological systems [1].

The three main ions (K^+, Na^+ and Cl^-) that participate in the generation of membrane potential, in cells in general, also perform other important actions in multiple cells, tissues and organs. In nerve fibers and muscle cells, the relationship between intracellular and extracellular potassium determines neuromuscular excitability [2]. Potassium is also involved in cardiac pacemaker activity and smooth muscle physiology and is fundamental for glycemic homeostasis [3]. Potassium exerts hepatic effects and integrates a negative feedback loop, which controls pancreatic insulin secretion, as well as the activity of the Na^+/K^+ pump ATPase, in cells in general. On the other hand, it also acts in protein synthesis and participates in enzymatic reactions [4]. Changes in the potassium gradient, typically resulting from changes in extracellular potassium, can be extremely important, both physiologically and clinically [5].

Ionic channels permeable to ions are present in living beings, including mammals. They are responsible for electrical transmission throughout the nervous system and participate in numerous physiological and biochemical processes, such as muscle contraction and the secretion of neurotransmitters and hormones. A notable feature of all living cells is the potential difference between intra- and extracellular fluids. This potential difference usually varies between 10 and 100 mV, with the interior of the cell being electronegative in relation to the outside [6]. The membrane potential is implicated in numerous cellular processes, such as ionic transport, cell signaling, electrical signaling in excitable cells, and other functions [7].

2. Patch-clamp technique

Patch clamp is a biophysical technique used for measuring the electrical activity of cells, including neurons and immune system cells. The cells can communicate with each other, generating electrical currents, and this process is referred to as bioelectrogenesis. This event is mediated by several types of inorganic molecules, ions and membrane proteins. When activated by a signaling molecule, membrane receptors activate a signal transduction pathway, which is an intracellular pathway. This signal transduction pathway involves the modification of the activity of intracellular proteins. This pathway is responsible for generating a cellular response producing communication with another cell, as observed for neuron-neuron, neuron-muscle cell and cell-cell communications [8, 9].

The use of new techniques and tools advanced in the fields of membrane biophysics and electrophysiology has allowed us to understand the structure and function of ion channels, the molecular basis of which is fundamental for the occurrence of electrophysiological phenomena. Other plasma membrane proteins function as transporters, allowing the flux of cationic (Na^+, K^+, Ca^{2+} or anionic (CL^-) ions during cellular communication. Therefore, when an isolated or connected cell possesses channels and receptors activated and functioning simultaneously, electrical recording is difficult; therefore, identifying and understanding these events is essential. The patch-clamp technique results in distinct configurations, such as the attached, whole-cell, inside-out, outside-out, and other configurations. Some configurations record the electrical activity of individual cells, others of an entire cell, and, in some cases, a network of cells. Therefore, the electrical activity of a single type of ion channel or neurotransmitter can be studied separately from that of many other channels in the same cell [10, 11].

3. Patch-clamp new applications

This volume describes a range of standard and novel methodological approaches used to determine ion channel function across different modalities [12]. The book contains fundamental concepts merged with detailed data providing background and context to the method for the nonexpert. The volume covers chapters with previously mentioned approaches, such as macropatch recordings, bilayer recordings, dynamic clamp, organotypic slice culture, combined in vivo patch clamp recording and optogenetics, and multielectrode arrays [13].

Patch-clamp configurations have not experienced any substantial innovations over the past few years. However, the use of this technique with other technologies that have advanced in the fields of optics and genetics has allowed the association of elements and the planning of more refined experimental tests. Thus, the combination of electrophysiology with multiphoton laser imaging and optogenetics is detailed at the molecular level in terms of the functional activity of single cells and whole circuits in tissues. These techniques can be used to evaluate the electrical properties of cells in 2D or 3D cultures or to record freely moving animals, providing incredible scientific insight [14–16].

Additionally, automated patch clamp systems enable rapid high-throughput electrophysiological screening that can test hundreds of compounds per hour. This highlights the relevance of this technique for industries seeking new selective prototypes for ion channels [17]. Cell excitability can also be measured via the use of voltage-sensitive dyes, and genetically encoded voltage indicators (GEVIs) can target ionic channels to the cells of interest [18]. In a similar manner, fluorescent calcium reporters can also be genetically encoded and allow the optical recording of action potentials in neurons [19, 20]. The incorporation of these technologies into cell-based assays and the development of automated high-content systems combining these optical tools with stem cell technologies represent exciting new phases in the era of drug discovery.

Author details

Robson Xavier Faria
Laboratory for Environmental Health Assessment and Promotion, Oswaldo Cruz Institute, Rio de Janeiro, Brazil

*Address all correspondence to: robson.xavier@gmail.com

References

[1] Schoffeniels E, Margineanu D. Molecular basis and thermodynamics of bioelectrogenesis. In: Topics in Molecular Organization and Engineering (MOOE, Volume 5). Dordrecht, Boston: Kluwer Academic Publishers; 1990

[2] Gutknecht J. The origin of bioelectrical potentials in plant and animal cells. American Zoologist. 1970;**10**:347-354

[3] Zheng H, Drumm BT, Zhu MH, YKE O'D, Baker SA, Perrino BA, et al. Na^+/ Ca^{2+} exchange and pacemaker activity of interstitial cells of Cajal. Frontiers in Physiology. 2020;**11**:230. DOI: 10.3389/fphys.2020.00230

[4] Palmer BF. Regulation of potassium homeostasis. Clinical Journal of the American Society of Nephrology. 2015;**10**(6):1050-1060. DOI: 10.2215/CJN.08580813

[5] Berend K, Hulsteijn LH, Gans TOB. Chloride: The queen of electrolytes? European Journal of Internal Medicine. 2012;**23**(3):203-211. DOI: 10.1016/j.ejim.2011.11.013

[6] Schofield Z, Meloni GN, Tran P, Zerfass C, Sena G, Hayashi Y, et al. Bioelectrical understanding and engineering of cell biology. Journal of the Royal Society Interface. 2020;**17**(166):20200013. DOI: 10.1098/rsif.2020.0013

[7] Abdul LK, Stacey M, Barrett-Jolley R. Emerging roles of the membrane potential: Action beyond the action potential. Frontiers in Physiology. 2018;**9**:1661. DOI: 10.3389/fphys.2018.01661

[8] Pei X, Volgushev M, Vidyasagar TR, Creutzfeldt OD. Whole cell recording and conductance measurements in cat visual cortex *in vivo*. Neuroreport. 1991;**2**:485-488. DOI: 10.1097/00001756-199108000-00019

[9] Margrie TW, Brecht M, Sakmann B. *In vivo*, low-resistance, whole-cell recordings from neurons in the anesthetized and awake mammalian brain. Pflügers Archiv: European Journal of Physiology. 2002;**444**:491-498. DOI: 10.1007/s00424-002-0831-z

[10] Brackenbury WJ. Patch Clamp Electrophysiology Methods and Protocols. Bioelectricity. 16 Jun 2021;**3**(2):154-155. DOI: 10.1089/bioe.2021.0015

[11] Verkhratsky A, Parpura V. History of electrophysiology and the patch clamp. Methods in Molecular Biology. 2014;**1183**(2):154-155. DOI: 10.1007/978-1-4939-1096-0_1

[12] Hodgkin AL, Huxley AF. A quantitative description of membrane current and its application to conduction and excitation in nerve. The Journal of Physiology. 1952;**117**:500-544

[13] Scanziani M, Häusser M. Electrophysiology in the age of light. Nature. 2009;**461**:930-939

[14] Boyden E, Zhang F, Bamberg E, Nagel G, Deisseroth K. Millisecond-timescale, genetically targeted optical control of neural activity. Nature Neuroscience. 2005;**8**(9):1263-1268

[15] Zhang F, Wang L, Brauner M, Liewald J, Kay K, Watzke N, et al. Multimodal fast optical interrogation of neural circuitry. Nature. 2007;**446**(7136):633-639

[16] Prakash R, Yizhar O, Grewe B, Ramakrishnan C, Wang N, Goshen I, et al. Two-photon optogenetic toolbox for fast inhibition, excitation and bistable modulation. Nature Methods. 2012;**9**(12):1171-1179

[17] Comley J. Outsourced Ion Channel Testing Trends. Technology Networks. 12 April 2017

[18] Hochbaum D, Zhao Y, Farhi S, Klapoetke N, Werley C, Kapoor V, et al. All-optical electrophysiology in mammalian neurons using engineered microbial rhodopsins. Nature Methods. 2014;**11**(8):825-833

[19] Chen T, Wardill T, Sun Y, Pulver S, Renninger S, Baohan A, et al. Ultrasensitive fluorescent proteins for imaging neuronal activity. Nature. 2013;**499**(7458):295-300

[20] Siegel M, Isacoff E. A genetically encoded optical probe of membrane voltage. Neuron. 1997;**19**(4):735-741

Chapter 2

Electrophysiology of Isolated Crayfish Neuron: Uncovered Neural Mechanisms of Eureka Moment

Toru Yazawa

Abstract

Our brain's neural network is complex. For studying the physiology of complex brain, sophisticated methods such as imaging and multielectrode recordings seem to be required. However, single-neuron level study is critical to understand the network functioning. While we conduct glass microelectrode experiments on the crustacean stretch receptor neuron, we found a phenomenon that may explain a flip-flop or bistable switching. The key mechanism is chloride ion–mediated GABAergic shunting action. In the present study, we argue that GABAergic synapses are key mechanisms for "switching" and we suggest a novel idea that can explain the mechanisms of sudden switching happening in the brain. We discuss how the mechanism is beneficial for the rapid brain response like "eureka moment."

Keywords: axon initial segment, CNS, crayfish, flip-flop, isolated neuron, GABAergic synapse, glass microelectrode, shunting inhibition, soma, switching

1. Introduction

Neural network of the central nervous system (CNS) is complex. Kuffler mentioned that "CNS is an assembly of cells that continually receives information, analyze it, perceives it, and make decision" ([1], fourth ed. Chapter 1, p. 4). Buzsaki wrote that "CNS is perpetually active and the brain's activity is generated from within, and perturbation of the default pattern by external inputs" ([2], p. 10). However, as Eccles stated that "it's taking a while to get final complete understanding of the brain" ([3], Preface, line 2). Even today, the brain and its working mechanisms are poorly understood.

Microelectrode method is apparently too simple to see the whole brain. Therefore, neurophysiologists use electroencephalogram recording, brain imaging, local field potential recordings, multisite probe recording [2], etc. for the purpose of studying the large-scale brain.

However, the glass microelectrode method is still useful, because only microelectrode can see details of electric current flows and accurate voltage changes occurring in the individual nerve cell. In the present study, we use glass-capillary microelectrode techniques to study the function of the nerve cell.

1.1 The nerve cell: From invertebrates to humans

Figure 1 shows that nerve cells of invertebrates have an "isolated" cell body (or soma) where the nucleus resides. The nucleus is a large organelle. It occupies larger space within the cell. Evolutionally, it is understandable that an obstructive oval structure (soma) derailed from main information-flow pathway as shown in **Figure 1**. Importantly, in **Figure 1**, one can see that any protrusion (dendrite) does not connect to soma membrane. Therefore, in **Figure 1**, soma is not involved in the path of action potential conduction. Information thus directly and firmly travels in the nerve cells. What about the case of vertebrate? (See **Figure 2**.)

Brown's textbook [8] shows five types of nerve cells in human brain. We follow him. **Figure 2** shows simplified diagrams: the dorsal root ganglion cell (A), sympathetic neuron (B), motoneuron (C), Purkinje cell (D), and pyramidal cell (E).

We have learned the general consensus (see Eccles [3] Figures 1–3): information signal arrives at the dendrites (black protrusions in **Figure 2B-E**), in the form of synaptic potentials. Then, information signal is integrated electrically during conducting through the dendrite-soma-axon (DSA) configuration. Finally, the

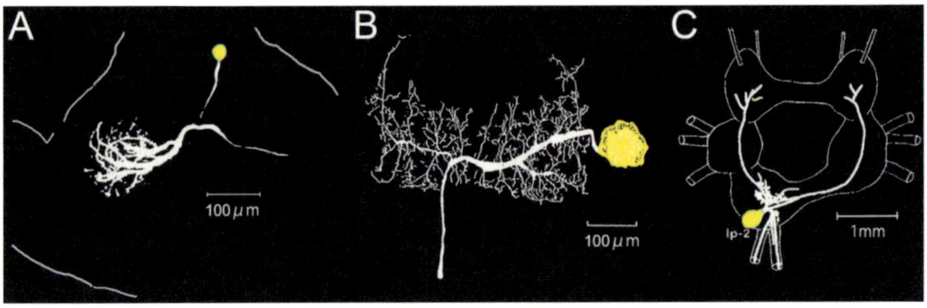

Figure 1.
Invertebrate nerve cells. Soma is colored in yellow. Note, soma has no protrusion, that is, the dendrite. Redrawing by the author: A, an insect nerve cell [4]. B, a crustacean nerve cell [5]. C, a molluscan nerve cell [6].

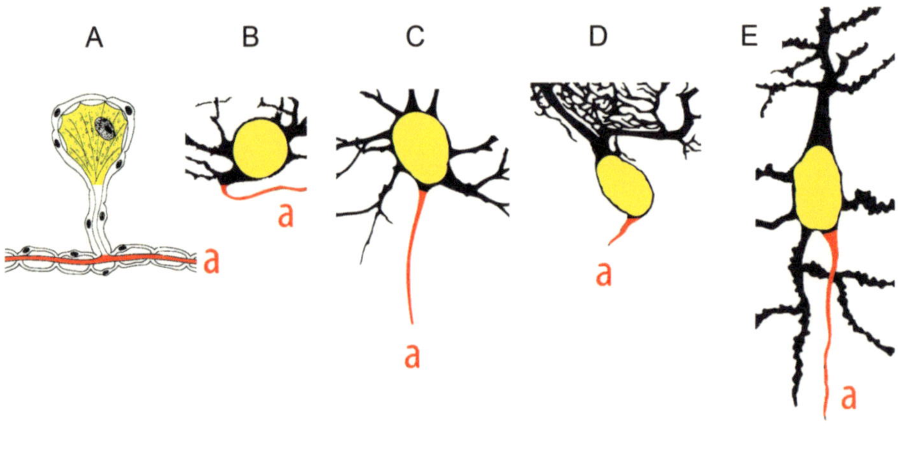

Figure 2.
Human nerve cells. Soma and its surroundings. Axon (a, red), soma (yellow), and dendrites (black). Schematic drawing by the author. A, a sensory neuron just outside spinal cord (see Matthews Figure 5–2 [7]). B, a sympathetic postganglionic neuron. C, a spinal alpha-motoneuron. D, a cerebellar Purkinje cell. E, a cerebral pyramidal cell. (See Brown [8] Figure 1.2).

neural output is transmitted through the axon in the form of action potentials (red colored in **Figure 2B-E**).

In **Figure 2A**, one can see the absence of "dendrite." These morphological and physiological properties are the same as those of invertebrate neurons shown in **Figure 1**. In this type of neurons, electrical information signal (i.e., action potential) does not pass through the cell body.

In **Figure 2B-E**, on the other hand, a large soma sits right on the middle of the path of information flow. The DSA configuration is common, and that is basic architecture of our brain cells.

When I affiliated electrophysiology department as an undergraduate student, I wondered: does the large spherical soma structure interfere with the accurate operation in the DSA-type neuron? Any textbook that I had [1–3, 7, 8] did not pay attention to the issue. We here categorize two types of neurons: Category A (**Figures 1** and **2A**) and Category B (**Figure 2B-E**).

Why one type neuron needs soma and the other type does not? Stuart et al. [9], who use microelectrodes, questioned similarly to us: how are synaptic inputs integrated in the intact brain? The debate is not settled yet. Microelectrode experiments might still be an important method to investigate the brain.

1.2 Microelectrode penetration into axon, soma, and dendrite: Orthodromic and antidromic action potential

Eccles mentioned that "impulse is a message traveling as a brief action potential (AP) in nerve" ([3], p. 238). And, therefore, microelectrode is one of the best ways for brain physiologists to investigate details of massage [9].

Here, we focus on the direction of message traveling in nerve. Ottoson [10] explained how AP (message) starts. We follow him in the present study (**Figure 3**):

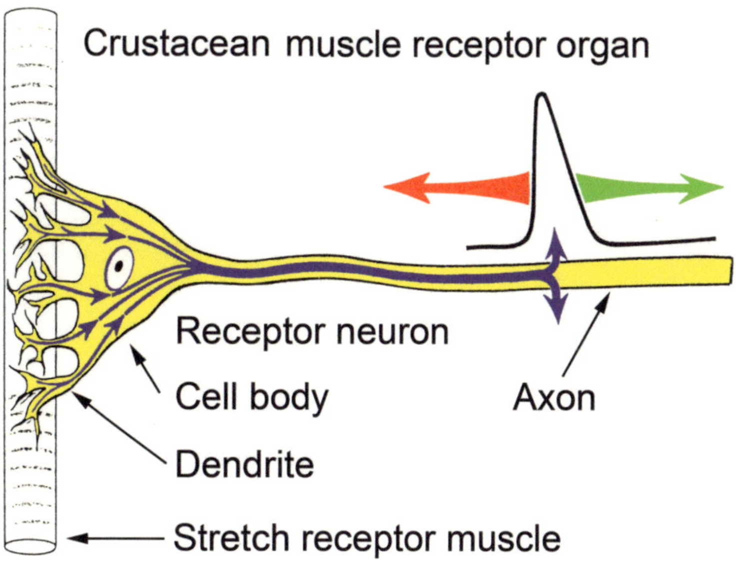

Figure 3.
Diagrammatic representation of impulse conduction in a neuron. The input current (blue arrows) electrically flows to the axon initial segment (AIS). An impulse starts at AIS. The impulse propagates in both directions: orthodromic (green arrow) and antidromic (red arrow). See [10, 11].

(a) The current produced in the dendrite flows into the axon. (b) The current passes out through the axonal membrane in the extracellular space at the "impulse initiation site" or "trigger zone." (c) The impulse, initiated at a distance of some hundred micrometers from the cell body (soma), propagates in a centrifugal direction along the axon as well as back into the soma toward the dendrites (see [10, 11]). We name the centrifugal AP and backward AP as orthodromic AP and antidromic AP, respectively, in the present study.

Antidromic AP conducting along the axon-soma-dendrite axis is always observable with the experimental conditions, as shown by Tauc [12], Eccles [3], and Stuart et al. [9, 13]. But the functional role of the antidromic AP is poorly understood, although Stuart et al. believe that it plays a greater role in information processing [9, 13].

Our aim is to study how a neuron relays the message to the other neuron. Until today, since functional significance of antidromic- and orthodromic-AP conduction within a single nerve cell is poorly understood, interaction between two nerve cells, such as interaction between principal neurons and interneurons, might not be as simple as we thought before.

1.3 Interaction between principal neuron and interneuron

Figure 4A shows that the surface of a principal neuron is covered by synaptic buttons. It is a highly packed state. Details were accomplished by means of serial electron micrographs of the spinal motoneuron (see Eccles [3], Figure 1–4). **Figure 4A** is a simple diagrammatic representation.

Figure 4B shows renowned drawing of a cerebellar basket cell (B). One can see that axon terminals encompass soma of the Purkinje cell. This perisomal network structure can be found in other brain cells: pyramidal neurons in cerebral cortex, pyramidal neurons in hippocampus, spinal motoneurons, etc.

It is known that there are many nerve terminals contacting with the principal neuron soma, and they are largely terminals of interneurons located in the vicinity. And furthermore, interneurons are immunohistochemically identified as parvalbumin-expressing neurons [15, 16].

Most importantly, parvalbumin (PV) neurons are largely GABAergic (meaning, gamma-amino butyric acid is the neuron's neurotransmitter substance) neurons.

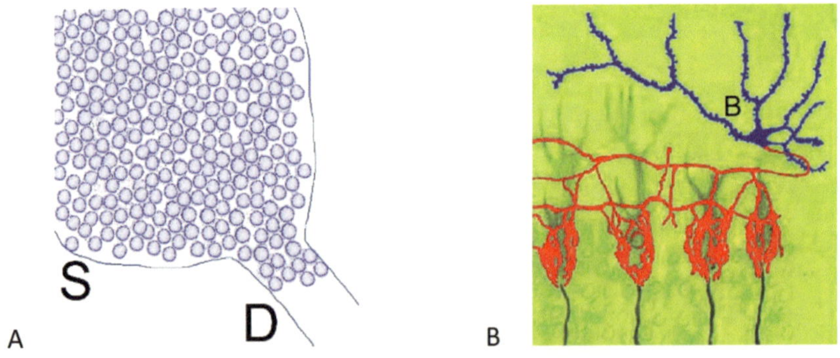

Figure 4.
A. Synapses on the surface of neuron. Diagrammatic representation of synaptic buttons made by the author. Soma area (S) and proximal portion of apical dendrite (D). B. Cerebellar basket cell (B). Soma and dendrites in blue, axon in red. Redrawing from original drawing by Santiago Ramon y Cajal [14]. Note, basket cell axon encompasses four large neurons, which are cerebellar Purkinje cells.

Nowadays, the "basket cells" are characterized as: they are (a) PV neurons, (b) GABAergic neurons, (c) inhibitory neurons, and (d) interneurons. Typical examples are reported in hippocampal, cerebral [15, 16], and cerebellar [14, 17] areas. Intriguingly, these areas are important areas that work for cognitive information processing.

Therefore, it is plausible that GABAergic neurons that innervate the soma have a canonical role. We consider that structure and function of GABAergic neurons would have common and universal characteristics in the brain.

1.4 Interaction between principal neuron and chandelier neuron

Chandelier (Cha) neuron makes synaptic contacts exclusively on the axonal initial segment (AIS) where action potential is generated [18]. This wiring completes during its development [19, 20], meaning that it is genetically determined. Cha neuron is "one of the most distinct types of interneurons, present in mammalian cortical circuits" [19]. We consider that Cha neuron might have a canonical role.

Using the publication [21], the author challenged to calculate how many pyramidal cells receive direct innervations from a single Cha neuron. The numbers were as follows: Cha 1, Cha 2, and Cha 3 innervate 61, 169, and 218 neurons, respectively (based on Figure 8 of [21]). So, a single Cha neuron innervates dozens of or hundreds of pyramidal neurons.

It is known that "abnormalities in chandelier cells have been linked to epilepsy, autism, and schizophrenia" [22]. With brain diseases, the patient should get a medical check of GABAergic cells' abnormalities [23]. In fact, evidences for GABA related brain neuropathology are increasingly accumulated [24]. It is understandable that substantial reduction of GABA availability in the brain, that is, loss of GABAergic neurons or impairment of GABAergic neurons, could exacerbate brain disease [25]. However, we still have limited knowledge about how GABAergic neurons shape our cognition. We still do not know accurately GABAergic neuron's role in the microcircuit. We should clarify what GABA terminals are doing at AIS and soma.

We report here membrane potential responses to GABA locally applied to the AIS-soma area. We use GABA-microinjection method and glass microelectrode recording method.

2. Materials and methods

2.1 Specimen

We used the isolated crayfish stretch receptor neuron as the model of the brain cells [1, 26]. This neuron has great utility: the morphology and physiology of the neuron have been well documented, and isolation/dissection procedure does not distract the neuron's original functioning [1, 10, 11].

Detailed methods are mentioned elsewhere [27]. Here, briefly, **Figure** 5 shows the procedure of how we dissect neurons. **Figures 6** and 7 show experimental setup, for membrane potential recording, axon stimulation, and microinjection of GABA. Oscilloscope (CRO) camera was used for recording, and electrophysiology data appeared on the CRO screen. Facilities at the laboratory of Kazuo Ikeda (deceased 2016), Neuroscience, City of Hope, Duarte, CA, USA, were used in 1985–1986. Crayfish *Procambarus clarkii* specimens were captured locally at Ontario, CA, USA (see [27]).

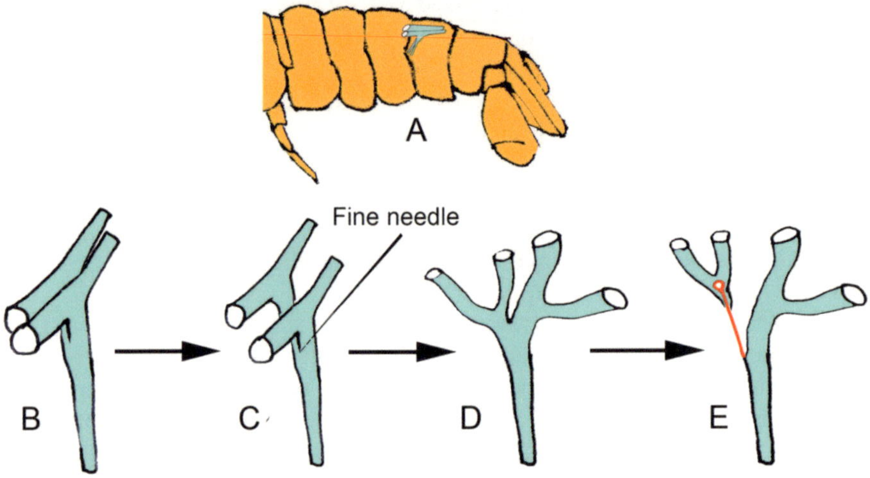

Figure 5.
Dissection of crayfish stretch receptor neurons. A, a location of the stretch receptor organ (colored in blue) in abdomen (brown). B, isolated muscle receptor organ, two organs sticking together; one neuron is slowly adapting sensory neuron, and the other fast adapting sensory neuron. C, gently separating each other by fine needle under the microscope. D, tearing more off by fine forceps. E, tearing further until surface of soma and axon (red in color) are exposed. The ideal conditions are as follows: first, microelectrode can measure resting membrane potentials; second, stretching of RM generates impulses in the neuron; and third, GABA application alters the value of membrane potentials. After confirming that GABA receptors localizing on soma and axon are physically exposed in a living state, we start experiments. (See [27], and ***Figures 6*** *and 7).*

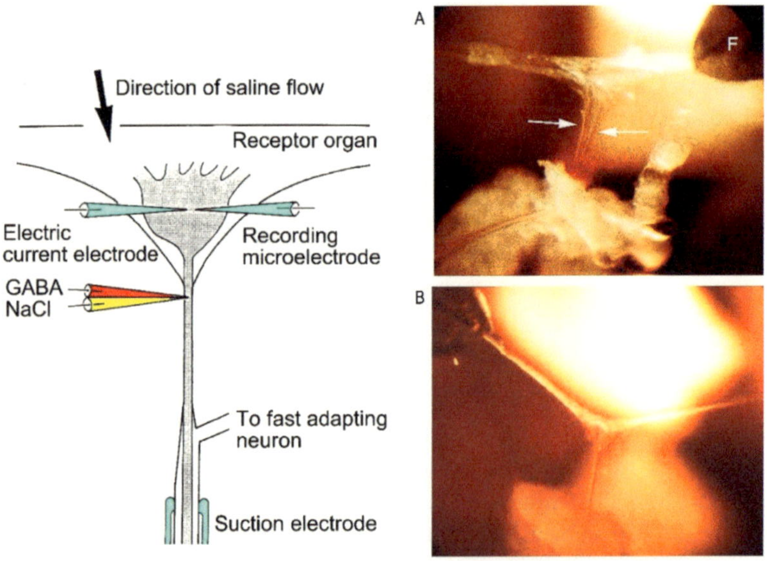

Figure 6.
Experimental setup of the isolated neuron specimen (see ***Figure*** *5 for dissection). Glass microelectrode for recording membrane potential (blue). Suction electrode for antidromic stimulation of axon of receptor neuron (blue). Double barrel electrode for electrophoretic application of GABA and NaCl (red and yellow, respectively), breaking current was passed when stopped using (see Ref. [27]). Inset pictures, taken through the dissection microscope by a digital camera. A, isolated muscle receptor organ. "F" in the photo denotes the forceps holding the receptor muscle. Note: two neuron's axons can be seen (two white arrows). B, GABA receptors of neuron are now exposed. A round cell body and a single axon can be seen. The surface of soma and axon is successfully exposed (see text and Ref. [27]).*

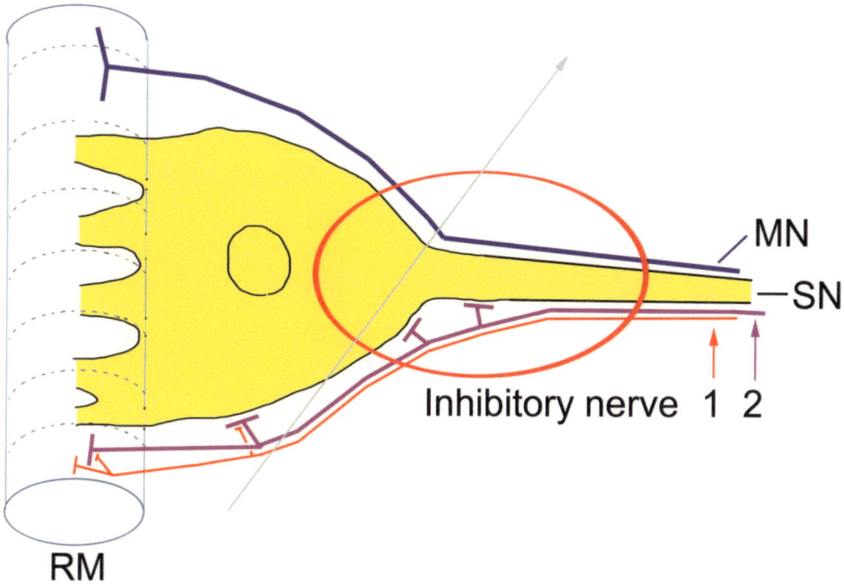

Figure 7.
*Diagrammatic representation of crayfish muscle receptor organ (or stretch receptor organ). See Kuffler's work in 1950s [1] for morphology and physiology. RM, receptor muscle. SN, sensory neuron (yellow). Extension of RM induces "generator potential" in sensory neuron, making SN discharge spikes [1]. MN, motoneuron axon innervating receptor muscle (blue). Inhibitory nerve 2 (thick nerve named in Ref. [1], pink) innervating sensory neuron's axon, soma, dendrite, and receptor muscle. Inhibitory nerve 1 (thin nerve named in Ref. [1], red) is not fully characterized but it innervates probably the receptor muscle, the sensory neuron, and the terminal of inhibitory nerve 2 (we have electrophysiological data of presynaptic inhibition: unpublished). Red elliptical shape indicates the area where GABA was locally applied. A gray arrow indicates the direction of flow of crayfish saline [27]. GABA was always applied to axon and soma-axon area in the present study (see the tip of GABA electrode in **Figure 6**).*

3. Results

3.1 GABA application to the axon: Flip-flop switching by GABA

Figure 8 shows the results of GABA application experiments. These experiments were reproducible [27]. GABA was applied with a current pulse (see **Figures 6** and 7 and Ref. [27]). Before the beginning of GABA experiments, the receptor muscle (RM in **Figure 7**) was pulled. The slowly adapting sensory neuron constantly fired impulses. As long as RM tension was steady, stable impulse-frequency discharges lasted for many hours in normal saline at about 15°C.

In **Figure 8A**, spontaneously occurring impulses are recorded at soma with a microelectrode. Impulse fires at 1.5 Hz in normal saline (impulses E-G). (Please refer to Kuffler's papers cited in Ref. [1] for details of this neuron's physiology.) The regular spiking indicates that excitatory information, that is, RM tension, constantly enters into the neuron, and the neuron is constantly sending out information downstream as train of impulses.

With GABA application, "sudden" change in frequency, from 1.5 Hz to 2.3 Hz, occurred (impulses H-W in yellow). Receptor muscle tension was unchanged throughout this experiment. This experiment clearly shows that GABA "switched" the discharging pattern. What we found is that GABA application revealed/uncovered that 2.3 Hz, instead of 1.5 Hz, is a genuine "frequency code" of the tension. This

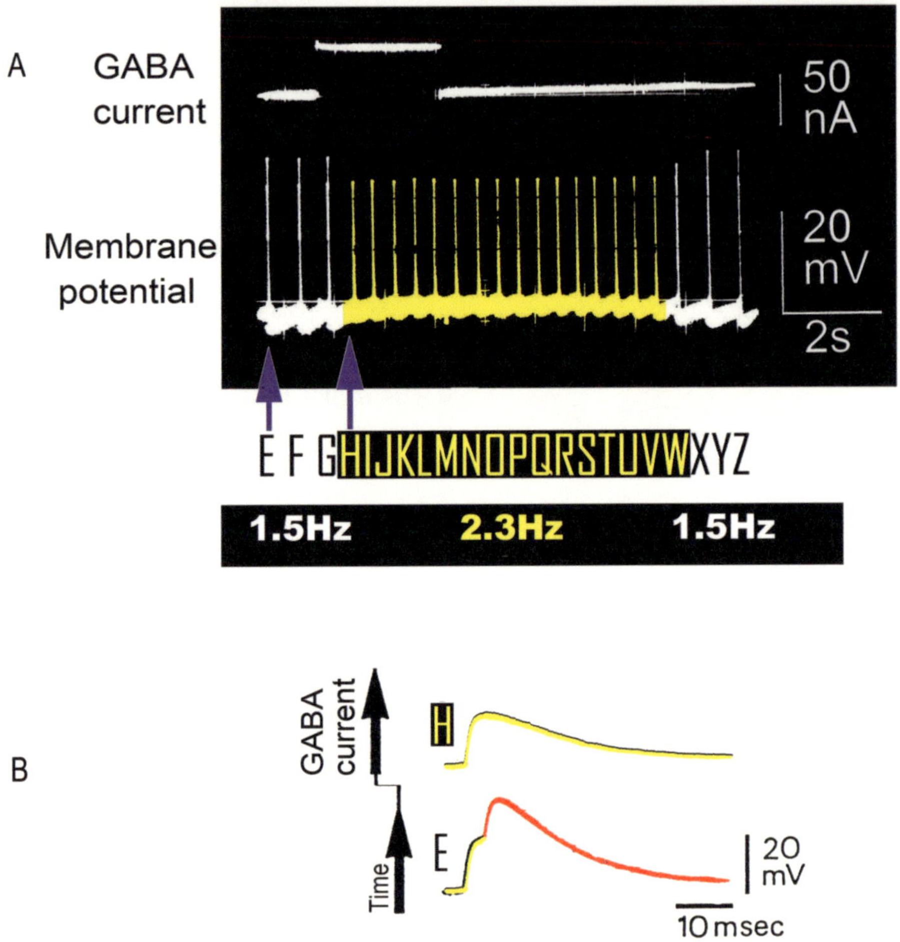

Figure 8.
GABA-injection experiment. A. Upper trace, GABA current monitor. Square current pulse, 50 nA for 2.5 second, used (see ***Figures 6*** *and* ***7****). Lower trace, membrane potential at soma. Slowly adapting sensory neuron. Spikes labeled as E–G (white) fire at 1.5 Hz. H–W spikes (yellow) fire at 2.3 Hz, which is caused by GABA. X–Z spikes (white) fire at 1.5 Hz. B. Fast sweep speed recording. Spikes E and H in A are shown in B with expanded time scale (the same cell but different experiment). Yellow spikes are spikes electrically transmitted to soma (IS spike). The IS spike started at the spike-generation site of the axon. Red portion of spike is regenerative S spike that occurred at the soma (see text). Note, in B, CRO camera operated with a "moving film" mode, so time goes from bottom to up. "E" spike, soma response in normal saline solution. "H" spike, soma response in GABA-containing saline solution. A and B, data from the same neuron specimen but from different experiment.*

experiment demonstrates that GABAergic synapses can switch from 1.5 Hz to 2.3 Hz, although receptor muscle tension was unchanged.

Figure 8A shows that switching ON from low rate to high rate occurs suddenly and switching OFF occurs suddenly again after the loss of GABA in the experimental chamber (impulses X-Z). Thus, it is clearly demonstrated that GABAergic interneurons, that is, chandelier neuron, can switch pyramidal neurons, that is, bistable ON/OFF or bistable high-frequency/low-frequency alternation.

GABA-induced sudden frequency increase is always accompanied by a sudden decrease of spike size, from larger size to smaller size. **Figure 8B** shows that the larger spike is composed of two components: (1) axon initial segment spike (IS spike) and (2) soma spike (S spike) (see Eccles and Tauc [3, 12]). Microelectrode at soma can see both IS spikes and S spikes. But if we record spikes at axon, we never observe whether or not the antidromic spikes invade into the soma-dendrite area. With the large-scale brain observation, such as electroencephalogram and/or brain imaging, we might not see antidromic invasion at all. Microelectrodes are still thus efficient for studying the brain.

Now, the question arises: why does the frequency decrease in "normal" saline? The answer is shown in **Figure 8B** (see also Ref. [27]). The causes of the reduction are invasion of the antidromic of impulse. Invasion of IS spike, that is, full-size AP generation at soma, always reduces the neuron's firing frequency. The reduction is distinctively large: it is approximately 35% in reduction from 2.3 Hz to 1.5 Hz.

Talented physiologists might say that GABA sometimes depolarizes neurons that may have excitatory effects. But, here, it should be noted that, GABA did not depolarize the neuron, as shown in **Figure 9**. Here, inhibitory postsynaptic potential (IPSP) shows "downward" swing. Here in this system, the inhibitory neurotransmitter is GABA, and in **Figure 9**, it is evident that the reversal potential of IPSP is more negative than the resting potential (please consult general neurophysiology textbook, such as Kuffler [1]; see also Eccles [3], Figures 3–6). So, the GABA's switching action is not due to GABA-induced depolarization but due to blocking of antidromic impulses.

Ottoson [10] revealed that the IS spike starts at axon (**Figure 3**). The IS spike is not full size all-or-none spike, because it is not the "regenerative" spike. The IS spike is an attenuated spike due to passive/electrotonic conduction (Kuffler [1]). Its size is variable due to damping caused by membrane capacitance and resistance. On the other hand, S spike is regenerative and full-size pike triggered by IS spike (see Kuffler [1] and Stuart et al. [9]). Therefore, we reach an important finding/conclusion that if IS spike fails to invade deep into the soma-dendrite area, it does not receive electric perturbations from the axon's spiking. In this specific condition, the neuron can endure accurate information processing; that is, the signal entering dendrites can travel safely (with high fidelity) to AIS without perturbations.

If the high-fidelity condition is established, the robust condition guarantees a noiseless environment for the information integration in the neuron (see also discussion in Ref. [27]). On the other hand, the state, allowing antidromic invasion, is not high-fidelity condition; that is, it is noisy state.

It looks like "inhibitory" GABAergic synapse "excites" the neuron (**Figure 8A**) as if the neuron received an excitatory information, that is, substance or stimulation.

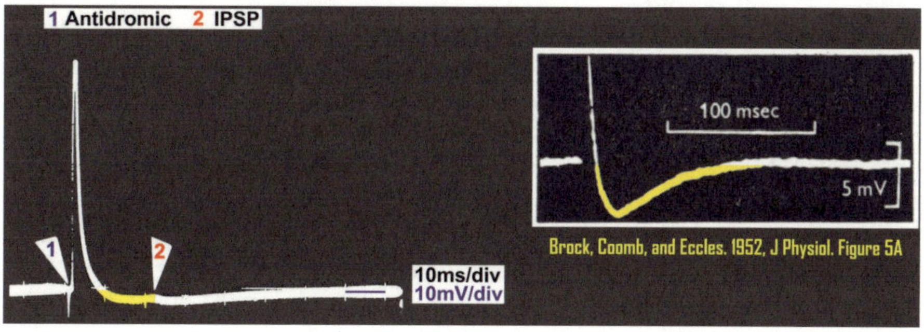

Figure 9.
Antidromic spikes and inhibitory postsynaptic potentials (IPSP) recorded at the soma of the slowly adapting neuron. Supper imposed three CRO traces. Paired stimulation applied: antidromic stimulation (1) and IPSP stimulation (2), 20 msec time difference between each other. See Ref. [27] for methods. Inset figure, redrawing from Ref. [28] to show what "afterpotentials" (colored in yellow) are, in general.

The enigmatic GABA effects are not enigmatic, because we demonstrated that excitation is caused by the blockade of antidromic spike invasion into soma and dendrite. GABA action in this case is not ordinal inhibition but "shunting" inhibition. It starts with a sudden increase in membrane conductance. And the membrane resistance becomes zero Ohm by shunting effect. This happens when the reversal potential of chloride ions is close to the resting potential (see neurophysiology textbook). Shunting is electrical phenomenon. Engineeringly speaking, the shunting phenomena can operate like the flip-flop switch, or switching between high and low logic levels. The electric phenomenon is evidently faster than chemical synaptic phenomenon. Electric phenomenon might be beneficial for the brain to do rapid response, like "eureka moment" and/or like hitting a fastball thrown at 100 mph (see below).

Now, we need to study neurophysiologically why antidromic invasion disturbs the accuracy (high fidelity) of information processing occurring in the soma-dendrite area (see **Figure 9**).

In experiments in **Figure 9**, we used totally different experimental setup than in **Figure 8**. Here, the receptor muscle (RM **Figure 7**) is not extended but slacken, so the neuron does not discharge spontaneously. The two receptor neurons are not separated by the dissection (see Methods; this specimen's state is the state in **Figure 5B**). Paired but independent electrical shocks evoked two independent responses in the soma of the slowly adapting neuron (**Figure 9**, marked as 1 and 2). (1) Antidromic spikes and (2) inhibitory postsynaptic potentials (IPSP) are captured simultaneously on the CRO screen (**Figure 9**). Methods: Two nerves were stimulated using suction electrodes; (1) slowly adapting neuron's nerve axon, antidromically, (2) inhibitory nerve 2 axon, orthodromically. (See Ref. [27] for methods in detail.)

Antidromic action potentials are all-or-none full-sized spikes. After the falling phase of the spike, one can recognize "afterpotentials" ("undershoot" or "underswing," partially colored in yellow, prior to IPSP swing). This afterpotential occurs because the potassium conductance during afterpotential is greater than at the resting potential (consult general neurophysiology textbooks). Typical "afterpotentials" are much more visible as shown in **Figure 9** inset (shown in yellow). Note, time-period length of "undershoot" potentials is as large as 100 msec. This is important for the neuron: with repetitive AP invasions into the soma, each AP brings a large hyperpolarizing afterpotential effect into the neuron. Thus, average membrane potentials are kept more negative than resting potential. So, average activity (frequency of discharge) of the neuron is suppressed due to net hyperpolarization, which is induced by antidromic AP invasions.

As shown in **Figure 9**, the undershoot potential (potassium and calcium ion's contribution) is slower in time course when comparing to the time course of the sharp spike potential (mainly sodium ion's contribution). Such a slow undershoot helps shift net value of membrane potential down toward much more negative level. As a result, antidromic invasion decreases the frequency of neuron's firings. In a general sense, antidromic invasion is unwelcome for the neuron's operation. But why does the neuron allow antidromic action potential invasion? The neuron might use invasion phenomenon efficiently. We hypothesize that the neuron efficiently uses the bistable characteristics, that is, noisy vs. not noisy, or logic high vs. logic low, like a flip-flop electric circuit. If so, antidromic invasion does not cause confusion in the nerve; rather, it might be more influential mechanism in subsecond brain switching processes. Indeed, it is known that GABA system's failure is exactly confusion, that is, mental diseases [24, 25].

4. Assessment

4.1 Shunting synapse

Back-propagating AP decreases discharge frequency of the neuron. In an experiment in **Figure 8**, the neuron fired at 1.5 Hz and 2.3 Hz during antidromic invasion mode and non–antidromic invasion mode, respectively. Here, the reduction rate is at 35%.

Previously, we already reported similar measures [27]. Following are these measures: 60, 39, 66, 60, 70, 71, 71, 59, 68, 57, 64, 57, 68, 61, 70, 77, and 78 (in percent). Average, minimum, and maximum values are respectively 64.5, 39, and 78 (in percent). In short, there is a large gap in firing frequency between the antidromic invasion mode and the non–antidromic invasion mode. The gap is roughly two times in frequency.

The most important fact is that the frequency suddenly increased about twofold within very short period, that is, in milliseconds. Imagine, in a quick change from 50 Hz to 100 Hz, it takes only 10 msec for the switching operation. The chemical excitatory synapse cannot afford such rapid switching operation.

The present studies also demonstrated that the microcircuit has two fixed-point attractors, for example, 50 Hz state and 100 Hz state. In other words, there are two steady levels in discharging pattern. This type of switching operation at GABAergic synapse has never been documented before in a vertebrate/human brain. Furthermore, the idea of bistability does not contradict with any previous data accumulated in brain science. We evaluate our results as followings.

4.1.1 Stepped change in spiking frequency by shunting

If a neuron fires at f (Hz) at "antidromic invasion mode," the genuine frequency code is as large as $2f$ (Hz). The healthy nerves generate spikes at a frequency roughly from 1 to 30 impulses per second (calculated from [29], p. 228 and [1], p. 330). The frequency reaches at most 190 Hz (calculated from Figure 17.6 [1]). Therefore, in general, normal nerve cells in the brain generate spike trains constantly firing at a frequency over the range from 1 to 190 Hz.

Aidley (see [29], p. 228) says that "information as the intensity of the sensory stimulus is carried as a frequency code: the greater the intensity of the stimulus, the higher the frequency of the all-or-nothing action potentials in the fiber." If the brain uses this code, the shunting can help quick message switching.

"Stepped change in spiking frequency by shunting" seems a plausible hypothesis. The present results do not contradict with the fact that impulse frequency can convey coded message and that stepped change helps rapid cognitive change in the brain.

4.1.2 Correlation between brain wave and shunting

In human brain evolution, emergence of shunting synapses might have been a great technical breakthrough for developing intelligent behavior of the brain. Neurons in the brain are known to work together, that is, synchronous firing of a group of neurons [2]. Shunting can control a group of neurons together by a quick operation like an electric switch. If an assembled group of neurons fires together in a phase for a certain period of time (e.g., for 300 msec [30]), sudden frequency jump (as $2f$ magnitude) can be detectable as characteristic brain wave, for example,

300 msec wave length oscillation in the extracellular space in the brain. That is why we can record larger enough amplitude and phase-locked wave, that is, electroencephalogram (EEG), resonating over a broad brain area. Shunting mechanisms do not contradict with those of EEG generation in the brain.

4.1.3 On/off switching operation and shunting

EEG electrodes would detect the synchronous sudden switching. However, both EEG and imaging (i.e., calcium imaging or functional magnetic resonance imaging) would not see details of action potential behaviors of individual neurons. Because of the single nerve recording methods, we discovered the shunting phenomena.

Crayfish neuron study revealed the existence of a neural quick switching (**Figure 8**). It is as if we are observing electric circuit behavior, that is, the logic low and logic high circuit. Such flip-flop operation is a highly possible mechanism for the brain. What brain's microcircuit needs is rapid switching when controlling cognitive tasks. A good example is a cognitive optical illusion, "Rubin's vase," which is ambiguous or bistable (i.e., reversing) two-dimensional diagrammatic image. In Rubin's vase, we see as "the black profiles of two people" or as "a white vase," but not both at once. Our brain might use alternative switching between: (1) colors black vs. white, (2) linguistically vase vs. face, (3) shape of vase vs. shape of face, etc. The switching can occur all together at once or independently. We hypothesize that the chandelier cells are able to do this task. The present results provide testable hypothesis that shunting enables rapid operation, that is, on/off switching.

4.1.4 Chloride ions and shunting

The ionic mechanisms regarding the shunting synapse are previously documented. At the shunting synapse, opening of chloride channel does not induce the movement of chloride ions. This is because the reversal potential for chloride ion (E_{cl}) is equal to the resting membrane potential (E_{rest}). How to keep $E_{cl} = E_{rest}$? It is known by the following.

To keep synapses' shunting mode, that is, maintaining $E_{cl} = E_{rest}$, it is known that there are two chloride ion transporters, that is, pump systems: (1) the chloride-extruding K^+- Cl^- cotransporter (KCC2) and (2) the $Na^+ - K^+ - 2Cl^-$ cotransporter (NKCC1). Theoretically, they can maintain intracellular concentration of chloride ions steady. (1) and (2) work together but independently. They can accumulate Cl^- within neurons, or they can maintain intracellular concentration of Cl^- steady. (See discussion by Ref. [27]).

In sum, the shunting switch works properly when E_{rest} is equal to E_{cl}. At such a synapse, electrical resistance of the synaptic membrane is null when chloride channels are open. In turn, electrical resistance is infinite when chloride channels are closed. So, there are two discrete states: null and infinite.

When at the null state, the electric current that generates action potential (driving force of action potential) is shunt. Then, action potentials do not conduct further ahead while chloride channels are open. This is the mechanism of shunting, interrupting conduction of antidromic impulses. It is clear that not sodium, potassium, and calcium ions, but chloride ions is the key ion for producing shunting effects due to the reversal potential theory. Previously recognized KCC2 and NKCC1 indeed support our working hypothesis derived from the results in **Figure 8**.

4.1.5 Functional significance of interruption of antidromic AP conduction

Assuming that the arbitrary diameter of axon and soma are five and fifty micrometers, respectively, and that sodium channels distribute evenly all over the membrane of the neuron, one can calculate that the soma's safety factor for generating action potential is "ten times" smaller than that of axon ([8], Chapter 4).

The lower safety factor guarantees that spike invasion into the soma is easier to interrupt by shunting, because of the reduced current density at the soma membrane, which is significantly lower than that at the axon, for example, ten times (see **Figure 10**).

The target of the basket cells is not fully clarified but the PV expressing, fast spiking, and GABAergic basket cells mainly innervate the soma of the target neuron (see [2], p. 69). We hypothesize a diversity of connection of basket cells: some of them selectively contact with (a) soma, (b) soma and proximal apical dendrite, and (c) soma and proximal basal dendrite (**Figure 10**). Therefore, higher command neurons can send command signal to the three [i.e., (a), (b), and (c)] portions simultaneously, independently, or sequentially (**Figure 11**).

Due to this diversity, the basket cell might help operate neatly or dynamically in human high-level cognitive tasks. For example, when we watch a tulip flower, the principal neuron (or neuron assembly) may integrate information: (a) a linguistic information "its name is tulip," (b) a shape information that "it's tulip shape," and (c) a color information "its color is pink" and so forth. During such information processing, the principal neuron (or neuron assembly) uses shunting synapses.

In our higher cognitive task, these varieties of switches enrich the brain's creative functioning. Humans are linguistic, artistic, musical, etc.

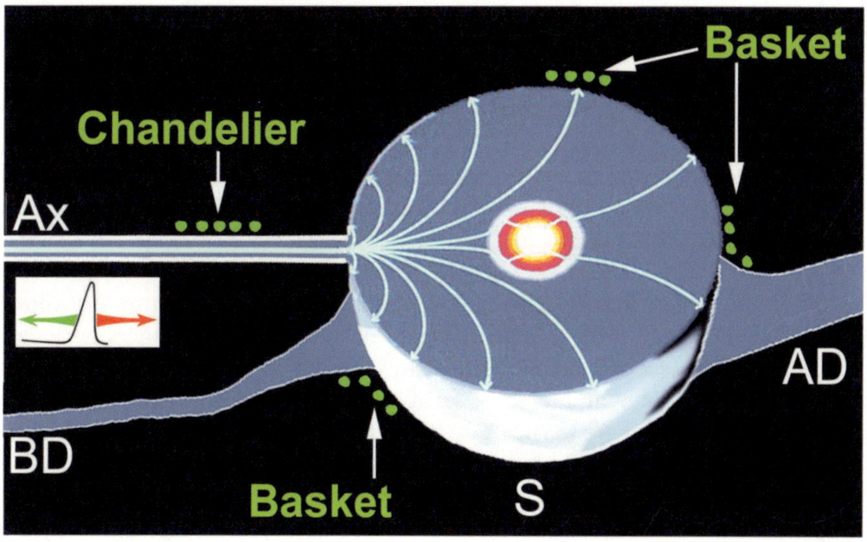

Figure 10.
Diagrammatic representation of back-propagating current flow (arrows in light blue color). An analogy from the seismic model by the author. Ax, axon. S, soma. AD, apical dendrite. BD, basal dendrite. Chandelier, green synaptic terminals that are made on the axon initial segment. Basket, green synaptic terminals that are made on the soma, the proximal portion of the apical dendrite, and the proximal portion of the basal dendrites. Inset, spike action potential started at the axon, traveling in orthodromic (green arrow) and antidromic (red arrow) directions.

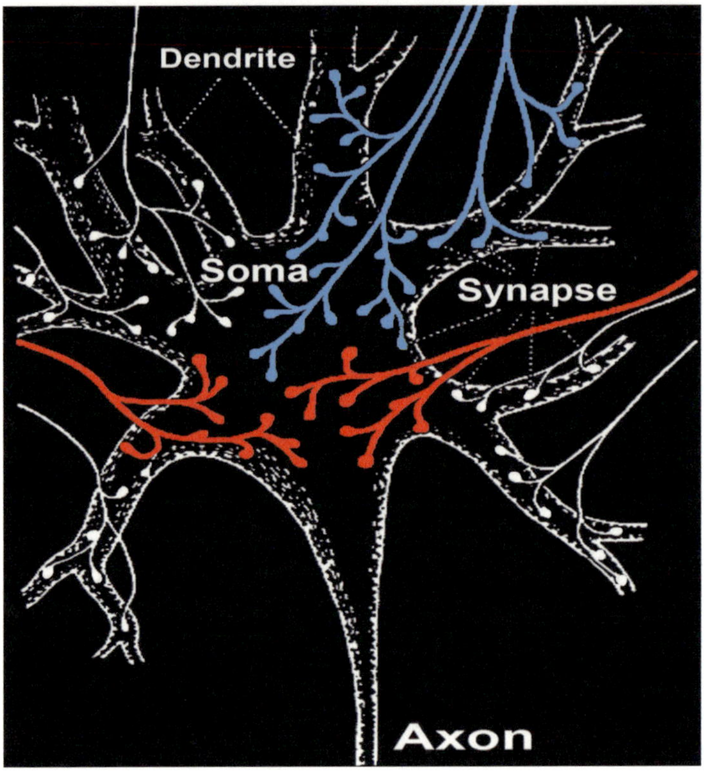

Figure 11.
Diagrammatic representation of a variety of inputs from GABAergic interneurons to the soma and dendrite. Interneurons (colored in blue, red, and white) represent axons from either a single neuron or assembled group of neurons. It is possible that "shunting" synapses operate synchronously, simultaneously, independently, or sequentially. Redrawn from Eccles [31], which is reconstructed from electron micrograph serial section.

4.1.6 Impulse train codes

Impulse train codes provide a succinct method to communicate with other neurons.

It seems impulse train carry messages. Would you decode the message from impulse train codes? There seems to be no consensus. However, we can explain it to a certain degree.

Figure 12 shows a diagrammatic representation of a neuron's output. No impulse in **Figure 12** (I) means that there is no information carried. For instance, impulse train codes could be decoded as following if an axon is a pain sensor axon: (1) an impulse train code depicts "an increased pain" or "pain exceeds threshold" or (2) an impulse train code tells where in the body the pain is generated right now: inside body, surface, or specific organ, and so forth.

Now, it is of interest to think about how higher brain uses impulse train code in the brain's tasks such as consciousness. **Figure 12** (N), (S), (R), and (E) may carry information. **Figure 12** (N) demonstrates a stable pattern, that is, uninterrupted regular impulse firing, code for which depicts "I am at online mode. I am listening/hearing/touching/watching" and so forth. Somewhere in the broad brain area, some neurons possibly use the "regular spiking pattern" code (N). Neurons are normally not silent but spiking at relatively low frequency.

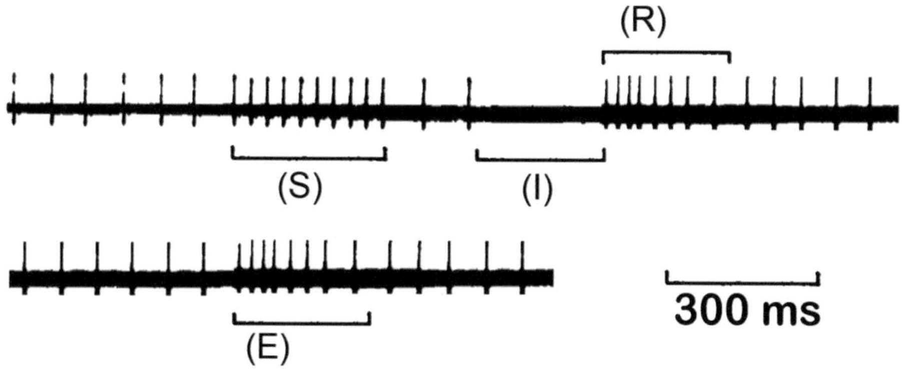

Figure 12.
Diagrammatic representation of train of impulses. (N), resting normal state, (S), shunting effects, for example, discharging at 30 Hz, gamma oscillation range [2, 40]. (I), inhibitory chemical synapse actions, for example, stopping discharge of the neuron. (R), rebound excitation after inhibition. (E), excitatory chemical synapse actions, for example, increasing discharge of the neuron (see text).

In digital machine–type information-exchange system, message transmission is based on all-or-none or 0/1 or HIGH/LOW system. We consider that (N) and (S) might be an unequivocal signal, depicting ON and Off, respectively. It is obvious that chemical synaptic excitation (E) and rebound excitation (R) are not unequivocal signals, although they are usable in pain sensor neuron.

We believe that we uncovered that, in the present study, 0/1 or HIGH-/LOW-type information transmission mechanism can exist in the brain.

5. Discussion

5.1 Evolution aspects

We categorize neurons into two types: Category A (**Figures 1** and **2A**) and Category B (**Figure 2B–E**). It is understandable that the Category A–type neurons need to simply transmit high-fidelity information without noise, such as what the spinal sensory neurons are doing. In turn, it is likely that Category B–type neurons (principal neurons and interneurons) evolved to engage in more complex neural integration function, for example, a eureka moment.

Category B–type neurons could efficiently operate in the human brain for decision-making: The following mechanisms are involved: (a) antidromic AP conduction contributes millisecond-level quick switching (**Figures 3**, **8**, **9**, and **11**), (b) axon initial segment GABAergic synapse might control dozens of principal neuron groups simultaneously such as the case of chandelier neuron (**Figures 8** and **10**), (c) GABAergic synapses on soma and proximal dendrites might control a variety of information-exchange tasks (**Figures 4**, **10**, and **11**).

It is clear that "simple sensory reflex operation such as a nociceptive spinal reflex" may not afford higher brain function. It seems that the human brain's basic evolution is accompanied by the innovation of Category B–type neurons.

Our findings indicate that Category B–type principal neurons cannot be simply viewed as the relay station delivering the dendrite input to the axon output.

5.2 Spiking and energy issues

Arguments in **Figure 12** indicate that spiking pattern is critical for conveying information. Rapid switching (see (N) and (S) in **Figure 12**) is able to (a) minimize logic redundancy, (b) reduce the circuit complexity, and thus (c) decrease energy consumption, because on/off shunting switch depends on electric switch that uses less energy than chemical neurotransmitter switch. At chemical synapses, a large chemical energy (i.e., ATP consumption) is required for (a) transmitter release, (b) transmitter reuptake, (c) transmitter breakdown, (d) keeping resting membrane potential steady by sodium-potassium pump, and so forth (see general text book). On the other hand, shunting switch simply obeys the Ohm's law.

Human brain uses only ~20 watts of power. (Rough calculation for example: 2400 food calorie = 2400Kcal. 2400Kcal/24 hr. = 27.8 cal/sec = 116.38 Jule/sec = 116 W. Brain's blood consumption = 20%. 20% x 112 W =23.3 W). Human brain is an energy-saving and high-performance machine compared to artificial intelligence/computer.

5.3 Microelectrode experiments

Microelectrode methods are widely used on vertebrate neuroscience studies [32–35]. In particular, Stuart and Sakmann have begun to use patch pipette in addition to sharp glass microelectrodes for investigating complex higher brain functioning [36]. To investigate higher brain functioning such as eureka moment dynamics, we believe we still need single-neuron level experiments in addition to imaging and other large-scale sophisticated technologies.

5.4 Spinal motor neurons may receive chandelier neuron recurrent synapse

The physiology of motor system is complex [1]. In fact, a motoneuron (MN) sends its axon to a variety of places, (a) the primary visual cortex, visual area 3, (b) the neurosecretory cells of the ventral hypothalamus, (c) other spinal MNs, and (d) the spinal Renshaw cells [37]. Renshaw cells are interneurons that innervate soma of MN. Renshaw cell contacts with MN with chloride ion–mediated synaptic mechanism. Renshaw cell's neurotransmitter is not GABA but glycine. Due to complex wiring of MN, the functional role of the Renshaw cells' function is not well characterized yet. The present study, however, strongly suggests that Renshaw cells have similar function as basket cells. This is a testable hypothesis.

"Motoneurons in the spinal cord have long been considered as the final common pathway of the nervous system, transmitting the neural impulses that are transduced into action" [37]. However, currently, "MN cannot be simply viewed as the last relay station delivering the command drive to muscles, but perform an active role in the generation and modulation of motor patterns" [37].

From the published photograph, we can observe that the spinal MN is likely to receive synapses from chandelier-type interneurons. There are six synaptic buttons on the MN's axon initial segment, although the author did not mentioned (see, [1], Figure 1.13 and [3], Figure 1–4). Thus, we hypothesize that chandelier-type interneurons and basket-type interneurons might work in the spinal MN system for performing very rapid tasks like baseball players do.

5.5 Spinal motor neurons: Baseball player's rapid response

Imagine, you are a baseball player who hits a fastball thrown at 150 km/h. "Good eyesight, concentration, experience, and reflexes; all of these are needed to hit a baseball at the Major League level [38]. The average hitter needs about 50 milliseconds to instinctively assess the speed and location of an incoming pitch. To squarely hit a fastball, everything has to be perfect. If you swing 7 milliseconds too early or 7 milliseconds too late, it's likely to be a foul ball" [38].

The neuron groups in the motor system including spinal, cerebral, and cerebellar neurons work together with subsecond quickness (50 milliseconds as Ref. [38] says). Therefore, we consider that the shunting switch is not a ridiculous idea in the execution of rapid performance.

6. Concluding remarks

Our original question was what is the difference between Category A–type and Category B–type neurons. The present experiments led us to find a novel and unexpected mechanism, that is, rapid neural switching.

We selected isolated neuron experiments using microelectrodes; otherwise, we could not find the mechanism. We would like to propose that the same mechanism might work in our brain at higher level brain functioning such as a eureka moment.

It is important that only matured adult brain experiences a eureka moment. An adult brain experiences a variety of experiences, and it stores/consolidates them in stocks of memory and/or neural network wiring. For instance, name of, shape of, and color of a flower are only stored in an adult's brain but not a baby's brain.

Indeed, we experience that our creative works require a creative mindset that needs experiences. We propose that almost all creative idea might spring from the matured brain using shunting switch at stochastic timing or decision timing. The timing could be observable as sharp wave ripple, a high-frequency wave in milliseconds wave length, that is, fast gamma bursts, 90–140 Hz in the time window of 50–150 ms [39]. Only a perfect brain can perform high-level cognitive operations such as a eureka moment.

The present study suggests that Cl^--mediated GABAergic inhibitory synapses on the axon initial segment and on the soma provide critical neural architecture for the flip-flop–type alternative switching. We would like to propose that this shunting synapses might work at a high-level brain functioning such as a "eureka moment" and "Rubin's vase," for example. This idea was derived from our glass microelectrode experiments on crustacean stretch receptor neurons. We really expect that vertebrate neuroscientists will confirm the shunting operation in vertebrate/human brain.

Definitely, the nerve axon conducts impulses. Impulses might carry information using a frequency code, which carries the intensity of stimulus, the strength of energy, the amount of information, etc. This code is also efficient for rapid flip-flop switch and thus efficient for high-level creativity of our brain. The frequency code of impulse trains and flip-flop mechanisms are the most plausible solution in the evolution of high-speed brain development. The working hypothesis is derived from crayfish neuron experiments. More studies are needed to show evidences for the same operation in human brain.

Acknowledgements

This work is supported from the JSPS grant 21 K12688 to TY.

Author details

Toru Yazawa
Tokyo Metropolitan University, Hachioji, Tokyo, Japan

*Address all correspondence to: tryazawa@gmail.com; yazawa-tohru@tmu.ac.jp

References

[1] Kuffler SW, Nicholls JG. In: Nicholls JG, Martin AR, Wallace BG, Fuchs PA, editors. From Neuron to Brain. 1975. 4th ed. Massachusetts, USA: Sinauer Sunderland; 2001

[2] Buzsaki G. Rhythms of the Brain. NY, USA: Oxford University Press; 2006

[3] Eccles JC. The Understanding of the Brain. NY, USA: McGraw-Hill; 1977

[4] Nishino H, Iwasaki M, Mizunami M. Pheromone detection by a pheromone emitter: A small sex pheromone–specific processing system in the female American cockroach. Chemical Senses. 2011;**36**:261-270

[5] Wine JJ. The structural basis of an innate behavioural pattern. The Journal of Experimental Biology. 1984;**112**:283-319

[6] Nishi T, Gotow T. A light-induced decrease of cyclic GMP is involved in the photoresponse of molluscan extraocular photoreceptors. Brain Research. 1988;**809**:325-336

[7] Matthews GG. Cellular Physiology of Nerve and Muscle. 3rd ed. Blackwell Science. Malden, MA, USA: Blackwell Science; 1998. ISBN 0632043547

[8] Brown AG. Nerve Cells and Nervous Systems. NY, USA. ISBN 3540196374: Springer-Verlag; 1991

[9] Stuart GJ, Spruston N. Dendritic integration: 60 years of progress. Nature Neuroscience. 2015;**18**:1713-1721

[10] Ottoson D. Physiology of the Nervous System. NY, USA. ISBN 0195204093: Oxford University Press; 1983

[11] Edwards C, Ottoson D. The site of impulse initiation in a nerve cell of a crustacean stretch receptor. The Journal of Physiology. 1958;**143**:138-148

[12] Tauc L. Site of origin and propagation of spike in the Giant neuron of Aplysia. The Journal of General Physiology. 1962;**45**:1077-1097

[13] Stuart G, Spruston N, Sakmann B, Häusse M. Action potential initiation and backpropagation in neurons of the mammalian CNS. Trends in Neurosciences. 1997;**20**:125-131

[14] Chedotal A, Richards LJ. Wiring the brain: The biology of neuronal guidance. Cold Spring Harbor Perspectives in Biology. 2010;**2**:a001917. Page, 1-17

[15] Hartig W, Brauer K, Bruckner G. Wisteria floribunda agglutinin-labelled nets surround parvalbumin-containing neurons. Neuro Reports. 1992;**3**:869-872

[16] Haunsø A, Celio MR, Margolis RK, Menoud P-A. Phosphacan immunoreactivity is associated with perineuronal nets around parvalbumin-expressing neurones. Brain Research. 1999;**834**:219-222

[17] Zhou J, Brown AM, Lackey EP, Arancillo M, Lin T, Sillitoe RV. Purkinje cell neurotransmission patterns cerebellar basket cells into zonal modules defined by distinct pinceau sizes. eLife. 2020;**9**:e55569. DOI: 10.7554/eLife.55569. Page, 1-31

[18] Somogyi P. A specific axo-axonal interneuron in the visual cortex of the rat. Brain Research. 1977;**136**:345-350

[19] Woodruff A, Yuste R. Of mice and men, and chandeliers. PlosBiology. 2008;**6**:e243

[20] Inan M, Blazquez-Llorca L, Merchan-Perez A, Anderson SA, DeFelipe J, Yuste R. Dense and overlapping innervation of pyramidal neurons by chandelier cells. The Journal of Neuroscience. 2013;**33**(5):1907-1914

[21] Blazquez-Llorca L, Woodruff A, Inan M, Anderson SA, Yuste R, DeFelipe J, et al. Spatial distribution of neurons innervated by chandelier cells. Brain Structure & Function. 2015;**220**:2817-2834

[22] Michalowski J. How Chandelier Cells Light up the Brain. Available from: https://www.cshl.edu/how-chandelier-cells-light-up-the-brain/ [Accessed: July 30, 2024]

[23] Gallo NB, Paul A, Van Aelst L. Shedding light on chandelier cell development, connectivity and contribution to neural disorders. Trends in Neurosciences. 2020;**43**(8):565-580. DOI: 10.1016/j.tins.2020.05.003

[24] Förster A, Model V, Gos T, Frodl T, Schiltz K, Dobrowolny H, et al. Reduced GABAergic neuropil and interneuron profiles in schizophrenia: Complementary analysis of disease course-related differences. Journal of Psychiatric Research. 2022;**145**:50-59

[25] Rana C, Mattis J. Pan-inhibitory hippocampal neuron ablation reveals insights into the role of interneurons in epileptogenesis. eNeuro. 2024;**11**(7):1-2. DOI: 10.1523/ENEURO.0229-24.2024

[26] Alexandrowicz JS. Muscle receptor organs in the abdomen of Homarus vulgaris and Palinurus vulgaris. Quarterly Journal of Microscopical Science. 1951;**92**:163-199

[27] Yazawa T. Isolated crayfish stretch receptor neuron electrophysiology may explain a longstanding mystery of human brain functioning: Eureka moment. In: Pertinent and Traditional Approaches Towards Fisher. London, UK: Intech Open; 2023. pp. 113-156. DOI: 10.5772/intechopen.109732

[28] Brock LG, Coombs JS, Eccles JC. The recording of potentials from motoneurons with an intracellular electrode. The Journal of Physiology. 1952;**117**:431-460

[29] Aidley D. The Physiology of Excitable Cells. 4th ed. Cambridge, UK: Cambridge University Press; 1998

[30] Magnabosco F, Hauk O. Decoding semantics from dynamic brain activation patterns: From trials to task in EEG/MEG tource tpace. eNeuro. 2024;**11**(3):1-14. DOI: 10.1523/ENEURO.0277-23.2023

[31] Eccles JC. Ionic mechanism of postsynaptic inhibition. Science. 1964;**145**:1140-1147

[32] Kole MHP, Stuart GJ. Signal processing in the axon initial segment. Neuron. 2012;**73**:235-247. DOI: 10.1016/j.neuron.2012.01.007

[33] Morris MM, Hao Z-Z, Berkowitz A. Electrophysiological activity of multifunctional and behaviorally specialized spinal neurons involved in swimming, scratching, and flexion reflex in turtles. eNeuro. 2024;**11**(7):1-16. DOI: 10.1523/ENEURO.0038-24.2024

[34] Higgs MH, Beckstead MJ. Impact of unitary synaptic inhibition on spike timing in ventral tegmental area dopamine neurons. eNeuro. 2024;**11**(7):1-19. DOI: 10.1523/ENEURO.0203-24.2024

[35] Stuart G, Spruston N. Determinants of voltage attenuation in neocortical pyramidal neuron dendrites. The Journal of Neuroscience. 1998;**18**(10):3501-3510

[36] Stuart GJ, Sakmann B. Active propagation of somatic action potentials into neocortical pyramidal cell dendrites. Nature. 1994;**367**:69-72

[37] Beato M, Bhumbra G. Synaptic projections of motoneurons within the spinal cord. In: Vertebrate motoneurons. Part of the Book Series: Advances in Neurobiology. Vol. 28. NY, USA: Springer; 2022. pp. 151-168

[38] Ramsay. Baseball: the physics of hitting a fastball. Brainscape academy. Available from: https://www.brainscape.com/academy/physics-of-hitting-baseball/#:~:text=A%20typical%20bat%20striking%20a%20ball%20can%20exert,100%20mph%E2%80%94the%20speed%20needed%20for%20a%20home%20run [Accessed: September 27, 2024]

[39] Buzsaki G, da Silva FL. High frequency oscillations in the intact brain. Progress in Neurobiology. 2012;**98**:241-249. DOI: 10.1016/j.pneurobio.2012.02.004

Chapter 3

Patch-Clamp: Studying the Passive and Active Properties of Excitable Membranes

Silvia Santillo

Abstract

The passive and the active electrical membrane properties are determinant keys in the signaling and information transfer between the excitable cells. They determine the input-output responses as well as the firing activity of neurons. This chapter will cover the description of the passive membrane parameters and their contribution in influencing the physiology of ion transport across the cell membrane. It will focus on protocol procedures to study and analyze both the passive and the active components using conventional whole-cell recordings. In particular, the voltage dependency and the kinetic parameters of the activation and inactivation curves of channels will be discussed in details.

Keywords: neuronal signaling, membrane equivalent circuit, passive electrical properties, voltage gating channel, activation/inactivation kinetics

1. Introduction

Neurons communicate with each other via electrical and chemical signals. During signaling, *i.e.* action potentials, synaptic potentials, or sensory generator potentials, the neuronal membrane voltage varies continuously causing changes in membrane potential of other neurons.

The patch-clamp electrophysiology, in its different configurations, is still considered a gold-standard method for studying the biophysical properties of excitable membrane and for understanding how they modulate the cell's electrical signaling. It allows to record, with high resolution, the transient variations of both membrane potential and ionic currents that occur during the neuronal signaling. It is a technique of choice also for studying the passive membrane properties of neurons, *i.e.* those properties which, although not involving voltage-dependent channels or receptors, are crucial in determining the rate at which a potential is conducted [1–6].

Usually, the membrane electrical properties can be modeled as an equivalent electrical circuit (RC) in which an ideal capacitor (C_m), *i.e.* the phospholipid layer, is in parallel with a resistor (R_m), *i.e.* the passive and active conductors. A battery is in series with the resistor and acts as electromotive force (EMF) with a voltage equal to the resting potential (**Figure 1**). The functioning of an RC circuit helps us understand

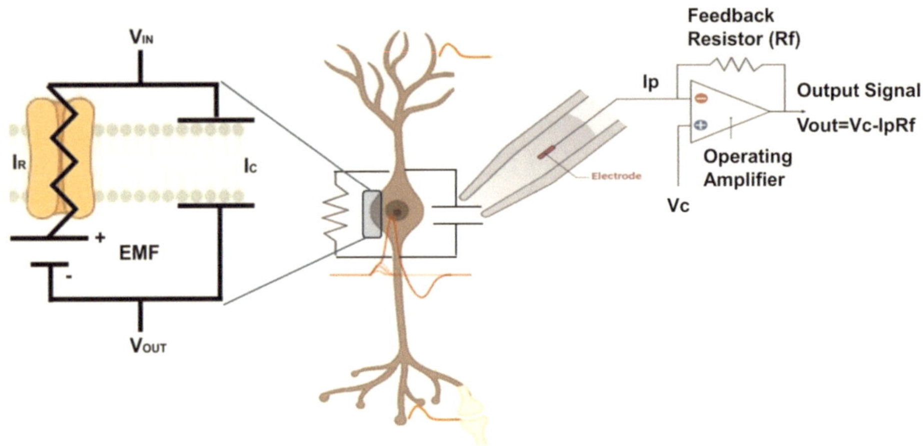

Figure 1.
The membrane equivalent electrical circuit. On the left the membrane is represented as an RC circuit in which capacitor is in parallel with a resistor. A battery in series with the resistor represents the electromotive force (EMF). Ion channels are represented as a single resistor element. On the right is depicted a simplify scheme of a patch-clamp amplifier in the voltage clamp mode. In this system the inverting input (-) of the amplifier, connected with the micropipette, is clamped to the command voltage (V_c) imposed at the non-inverting one (+). To ensure this clamp pipette currents (I_p), flowing through a feedback resistor (R_f), will be proportional to the output voltage of the amplifier, i.e., the difference potential needed to shift the membrane potential to the command potential. Created with BioRender.com.

how the capacitive component influences the rate of change of the membrane potential.

Initially, the voltage across the resistor and capacitor are equal as the resistance and capacitance of the membrane are in parallel because they are linked by low resistance paths, *i.e.* the horizontal line in the circuit diagram. After an instantaneous change of voltage, a large current will flow across the capacitor (I_C) while no current will flow through the resistor (I_R). Consequently, the total current flowing across the membrane will be represented by the only capacitive component ($I_m = I_C$) and expressed by the equation:

$$I_C = C_m \frac{dV}{dt} \tag{1}$$

This instantaneous current will induce a variation of charges across the capacitor and indeed of potential, as the voltage is proportional to the charge stored on the capacitor ($dV = dQ/C$). As a result, the membrane potential variation will become more positive and begin to drive current across the membrane resistance. The more current will flow through the resistor (I_R), the less across the capacitor as I_C plus I_R is a constant ($I_m = I_C + I_R$). This explains why the membrane potential rises more slowly and reaches its final value only when all the membrane current has flowed through the resistor that from Ohm's law is given by the equation:

$$I_R = \frac{V_m}{R_m} \tag{2}$$

Once the stimulus is turned off, the membrane capacitive currents dissipate through the only resistive component ($I_m = I_R$) inducing the membrane potential to return to its initial value. We can say that in response to a step input the membrane potential variations are indeed influenced at the initial slope only by the membrane capacitive component while at the final slope by the membrane resistive component. The equations that define the value of V_m, moment by moment, during the charging and discharging phase of the membrane are respectively the following:

$$\Delta V_m(t) = I_m R\left(1 - e^{\frac{-t}{\tau}}\right) \tag{3}$$

$$\Delta V_m(t) = I_m R\left(e^{\frac{-t}{\tau}}\right) \tag{4}$$

where τ is the time needed for the rising voltage to reach the 63% of the final value or alternatively the time required for the decaying voltage to reach the 36% of its initial value.

2. Membrane passive properties

The membrane passive properties are referred to those properties that do not involve voltage-dependent channels or metabotropic and ionotropic receptors but at least only ion leakage channels and pumps. The three main passive membrane properties are the capacitance, the input membrane resistance, and the resting membrane potential. They are interconnected with each other and influence the synaptic output and integration [7–9], the neuronal excitability, and the speed of signal propagation [10].

2.1 Membrane capacitance

The membrane can be considered as a capacitor that attracts charges, the negative and positive ions contained in the intracellular and extracellular media. The amount of charge that can be separated across the membrane is directly proportional to V_m, Eq. (1) and at a fixed membrane voltage, results proportional to the area of the membrane, *i.e.* the bigger the membrane is, the more charge it stores. It is defined by the formula:

$$C_{Farad} = \frac{Q_{Coulomb}}{V_{Volt}} = \epsilon_0 \times \frac{A}{d} \tag{5}$$

where ϵ_0 is the vacuum dielectric constant, the permittivity of free space $8.85 * 10^{-12} F^* m^{-1}$ (or the relative dielectric constant, an adimensional value expressed as the ratio of the dielectric permittivity of the material respect to the vacuum), (A) is the membrane surface, and (d) is the distance between the layers. Assuming a specific membrane capacitance of $1.0 \mu F/cm^2$ $(1pF/100\mu m^2)$ [10–13] it is possible from Eq. (5) extrapolate the membrane surface value (A).

The capacitive component of the membrane (C_m) determines the rate (τ) at which the membrane potential responds to the flow of ionic currents. It is a crucial parameter for the ability of the neuron to integrate the synaptic inputs it receives. In fact, playing on their temporal and spatial coincidence (temporal and spatial summation) [14] it affects the time course (τ) of electrical signals and determines how fast the neuron membrane propagates the signals defying the neuronal excitability [15].

$$\tau = R_m C_m \tag{6}$$

The higher the capacitance, the higher the τ and the longer time the membrane will take to reach its final value.

The response time (τ) also determines the cutoff frequency (fc) for signal filtering, *i.e.* the frequency upper boundary beyond which the filter will not pass a signal

$$f_c = \frac{1}{2\pi\tau} \tag{7}$$

The membrane, in fact, behaving like a parallel RC circuit acts as a low pass filter by removing the high frequencies and leaving a slow response to injected currents. It means that a cell capacitance of $10pF$ and a series resistance of $10M\Omega$ ($\tau = 100\mu s$) will filter the membrane currents with an effective f_c of about $1.5KHz$ (about $700\mu s$).

The more common method to estimate the C_m is a voltage clamp based approach [16, 17]. After a brief voltage step an initial current transient will be generated and charge the capacitor. Fitting by an exponential the decaying phase of the current response to the voltage step, we can extract τ when whole-cell capacitance and series resistance are not corrected (**Figure 2A**). The membrane resistance can be instead calculated by the Ohm's law (Eq. (2)) extrapolating from the exponential curve the peak current, which results immediately after starting the voltage step, and finally, the C_m value can be obtained from Eq. (6). Capacitance can be also derived determining the area under the transient of the current trace as it is proportional to the integral of the charging transient. Alternatively, largely congruent values can be also

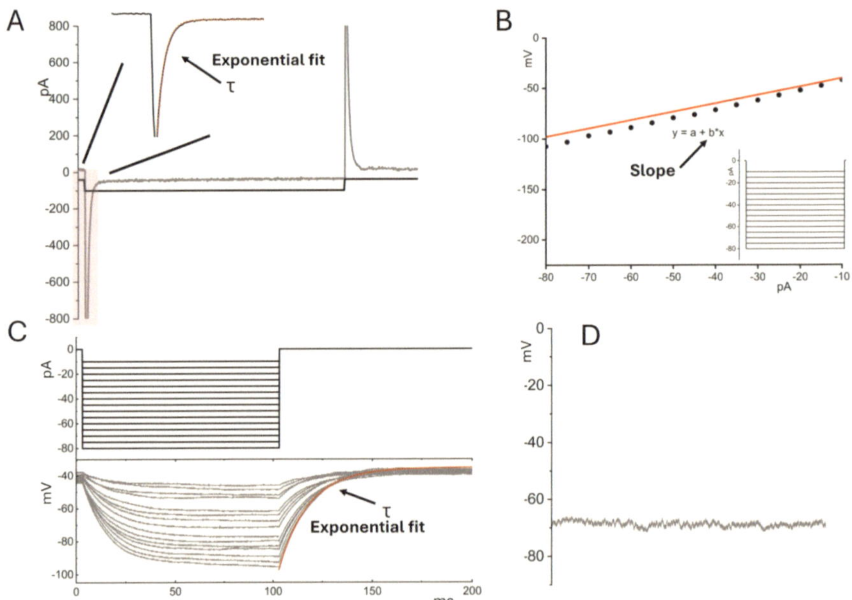

Figure 2.
Membrane passive properties. (A) The hyperpolarizing voltage step (black line) and the exponential fitting (red line) of the current decaying phase (gray line); (B) V/I relationship of voltage responses to hyperpolarizing current injections and the linear fitting (red line) to extrapolate the R_{INPUT}, the angular coefficient in a regression model (slope); (C) the temporal step protocol to calculate R_{INPUT}, C_m, and τ by means of a single experiment; (D) a representative RPM trace recording in CC mode.

obtained by applying the continuous square wave protocol *Membrane test* implemented in *pClamp*.

2.2 Membrane input resistance

The membrane input resistance (R_{INPUT}) is a passive membrane property that directly depends on the capacitance of the cell. The greater is the cell surface and the greater is the number of ohmic channels which, allowing the passage of current, represent a limiting factor on the excitability by influencing the timing of the voltage response to a stimulus.

R_{INPUT} also determines the magnitude of the depolarization that occurs in response to an injection of current. Consequently, between two neurons receiving an identical input current, the neuron that possesses the higher input resistance will result in a greater variation in membrane potential and in an increase of its excitability. This is because, according to (Eq. (2)), a given depolarizing current will produce a greater voltage increase with a greater chance of evoking an action potential [18].

A mandatory requirement to calculate the R_{INPUT} is that the membrane resistance does not change; indeed, hyperpolarizing voltage steps of small amplitude should be applied to minimize membrane conductance activation. The R_{INPUT} value is determined in the current-clamp mode by measuring the voltage responses to $-5pA$ current injections starting at a holding current of $0pA$. It is estimated as the slope of a linear fit (the angular coefficient in a regression model) of the voltage membrane values measured at steady state (when all of the current flows through the resistor) plotted against the current steps (V/I relationship, **Figure 2B**).

Alternatively, applying a protocol that, after the hyperpolarizing step, returns for a longer time to the initial current value allows R_{INPUT}, C_m, and τ to be extrapolated from the membrane discharge phase inside the same experiment (**Figure 2C**).

2.3 Resting membrane potential

The resting membrane potential (*RMP*) is defined as the difference between the inside and the outside of the cell membrane. It depends on the different intracellular and extracellular concentrations of ions, mainly potassium, flowing through the ohmic channels, their permeabilities (*P*), and pumps (Na/K-ATPase, NCX), which, restoring the appropriate balance of ions, maintain the membrane potential to its resting value. It is formalized in the Goldman-Hodgkin-Katz equation

$$Vm = \frac{RT}{F} ln \left(\frac{P_{K^+}[K^+{}_{OUT}] + P_{Na^+}[Na^+{}_{OUT}] + P_{Cl^-}[Cl^-{}_{IN}]}{P_{K^+}[K^+{}_{IN}] + P_{Na^+}[Na^+{}_{IN}] + P_{Cl^-}[Cl^-{}_{OUT}]} \right) \tag{8}$$

where each ion concentration is weighted by the factor *P*, which refers to how easily an ion crosses the membrane, indeed is directly proportional to the number of open channels, and is independent of ion concentration. As the ohmic channels are essentially permeable to potassium ion, the membrane permeability for this ion will be greater. Obviously, the bigger the permeability, the more that ion will influence the resting membrane value. This is why the resting membrane potential is close to the K^+ equilibrium potential, *i.e.* the membrane potential value at which the electrical field, developed across the membrane for the ion flow, perfectly balances the concentration gradient causing ions to move randomly across the membrane with a zero net flux.

The E_X is expressed by the Nerst equation:

$$E_X = \frac{RT}{zF} \, ln \, \frac{[X]_o}{[X]_i} \tag{9}$$

where the only variables are the external and internal ion concentrations, which, together with z, determine the sign of the equilibrium potential.

Membrane potential changes are crucial in the activation of membrane ionic conductances as in excitable cells ion channels are voltage-dependent with kinetics restricted to a range of values.

Even at the synaptic level, the flow of ions through some receptors depends on the membrane potential. NMDA receptors at the resting potential are blocked by Mg^{++} and small depolarizations can increase the likelihood of Mg^{++}-unblocking and facilitate the activation of receptor conductance as soon as it binds to glutammate [7, 8, 19, 20].

The *RPM* and all the spontaneously potential changes can be investigated in whole cell current clamp mode (CC), under gap free protocols without applying any holding current so that the membrane potential is free to vary (**Figure 2D**). Alternatively, it is possible to register the *RPM* in cell-attached mode as soon as the seal resistance increases by several orders of magnitude [21, 22]. It is important to remember that at the *RPM* some current is flowing through the membrane. Therefore, if we want to perform experiments at resting potential, the appropriate setting holding current will not be zero but the needed current to set the membrane at the resting potential.

3. Excitable membrane active properties

Neuronal membranes contain ligand- and voltage-dependent channels that allow specific ions to pass through them [23, 24]. The driving force for these ions is determined by the deviation of the membrane voltage from the equilibrium potential for that ion (E_X) and by the different ionic concentration existing at the two membrane sides. It represents the electromotive force (EMF) depicted in **Figure 1** as a battery in series with the resistor.

As ion flow is limited by the membrane resistance, R_m, Eq. (2) can be written as

$$I_X = \frac{E_m - E_X}{R_m} \tag{10}$$

where E_m-E_X is the driving force for the ion X and R_m the membrane resistance in which channels are in the closed state. Indeed, the resulting current is inversely proportional to R_m, *i.e.* the fewer channels are open, the smaller current will flow.

However, in electrophysiology the current flow across the membrane is represented as the reciprocal of the resistance or conductance (G, expressed in Siemens), which is a biophysical property of an ion channel. Then, Eq. (10) becomes

$$I_X = G_X(E_m - E_X) \tag{11}$$

where G_X is now simply represented by the sum of the individual conductances as channels are considered parallel capacitors.

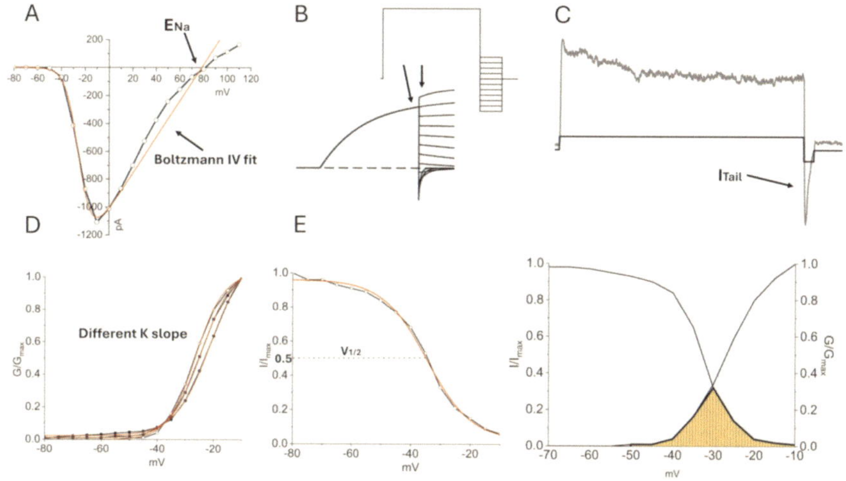

Figure 3.
Membrane active properties. (A) The Na^+ I/V relationship fitted to the transformed Boltzmann function (red line) and showing the E_{Na} within the range of the active conductances; (B) K^+ tail protocol and I_{Tail} to find the reversal potential when it lies in a membrane voltage range where conductances are not active; (C) protocol eliciting a rapid onset of hERG inactivation at depolarized potentials, with a fast inactivation recovery at the hyperpolarizing step; tail current is appreciable thanks to the slow channel deactivation; (D) activation kinetic curves shaping different slopes whose K values can be extrapolated by Boltzmann fitting (red lines); (E) inactivation kinetic curve showing the $V_{1/2}$ extrapolated using Boltzmann fitting; (F) the "window" of voltage resulting by overlapping the activation and inactivation curves.

The current values obtained by applying a voltage steps protocol when plotted in a current–voltage relationship (I/V curve) provides the voltage range activation and, for some ions (*e.g.* E_{Na}), the equilibrium potential (**Figure 3A**). Extrapolation of the linear part of the I/V relation or application of a tail current protocol are useful methods to find the reversal potential when its value is at a higher voltage membrane not included into the step protocol (*e.g.* E_{Ca}) or when the reversal potential lies in a membrane voltage range where channels are normally closed and conductances are not active (*e.g.* E_K). In the latter case, by activating a maximum current and then bringing the potential to a series of values close to the respective equilibrium potential, we will obtain a tail current, whose amplitude and direction, depending on the driving force, will indicate the value of the Nerst potential (**Figure 3B**).

Tail currents, I_{TAIL}, are also useful to study peculiar kinetic mechanisms like those of hERG channels. The gating of these channels is, in fact, characterized by a rapid onset of inactivation at depolarized potentials. Their fast recovery from inactivation combined with a slow channel deactivation induces a resurgent current when membranes repolarize. As I_{TAIL} depend on the inactivation characteristics of the channels, it is possible, by applying opportune protocols, to obtain this resurgent current as a function of recovery of inactivation (**Figure 3C**, [25]).

3.1 Activation and inactivation kinetics

The voltage-gated ion channel activity depends on both activation and inactivation kinetics, *i.e.* transitions from the resting state to the open and inactivation ones. The shifting of these curves toward lower or higher voltages can induce fundamental changes in the voltage dynamics of ion channels [25–29].

To study Na^+ activation curves, conductances (G) are calculated by Eq. (11) in which I_{Na} arises by means of a series of conditioning depolarizations, *e.g.* stepping the membrane to potentials ranging from -80 to $0mV$ by $5mV$ increments. The conductivity values normalized to the maximal conductance, G/G_{max}, will represent the percentage of maximum conductance, *i.e.* the fraction of activated channels as a function of the membrane potential (**Figure 3D**) [30, 31].

Unlike sodium channels, which undergo to a rapid inactivation, the activation kinetics of K^+ channels can be also evaluated by a two-pulse protocol consisting of a series of depolarizing voltage steps that return, at the end of depolarization, to a membrane value at which channels are in closed state. The steady state current at the end of chosen depolarizing voltages, or the initial amplitude of the tails, (see arrows in **Figure 3B**) can be used to generate the activation curves as they reflect the number of open channels at the tested potentials *i.e.* the channel open probability.

For obtaining the inactivation curves, I/I_{max} (**Figure 3E**), conditioning prepulses, *i.e.* potential prepulses ranging from -80 to $0mV$, are followed by a test pulse to obtain a maximum current, *e.g.* -10mV for Na^+ and $+40mV$ for K^+ currents [32]. The currents amplitude generated at the test pulse should be normalized to the maximum amplitude current value obtained by the conditioning pulses and will represent the unavailable channels probability at each membrane potential value of the test pulses, as function of the conditioning pulses (the protocol is quite similar to that depicted in **Figure 3B**).

By comparing the two activation and inactivation relationships (**Figure 3F**), the presence of an overlapping area between the two curves will indicate that a continuous current is present in that range of membrane potential values. This "window" of voltage represents the probability density that a few channels could be either in the open activated state or in the closed inactivated state. It means that the non-inactivated channels have a small likelihood of spontaneously opening at those voltages. The integral of this *window* is physiologically significant as it reveals the level of synchronization of the channels and therefore the efficiency of the membrane to respond to stimuli.

The mathematical fitting of IV and activation/inactivation curves provides many additional informations about the kinetic of the channel gating as it is possible to extrapolate for each x value of the fitted curves the respective y value and *vice versa*.

The IV data of sodium currents are generally fitted to the transformed Boltzmann function [30, 31] (red line in **Figure 3A**)

$$I = G_{max} \frac{(V_m - V_{Na})}{1 + \exp^{\left(V_m - V_{\frac{1}{2}}\right)/K}} \tag{12}$$

for which it is possible extrapolate the midpoint of activation ($V_{1/2}$), *i.e.* the membrane value at which the 50% of ion channels result activated. V_m and V_{Na} in the equation represent the test potential and the sodium reversal potential while the G_{max} represents the maximal conductance obtained at the depolarizing voltage step of sufficient magnitude to elicit maximum Na^+ conductivity and calculated by using the following equation

$$G_{max} = \frac{I_{max}}{V_m - V_{Na}} \tag{13}$$

The Na^+ activation curves can be fitted to the following Boltzmann function:

$$G/G_{max} = G_{max} + \frac{G_{min} - G_{max}}{1 + \exp^{\left(V_m - V_{\frac{1}{2}}\right)/K}} \tag{14}$$

where G_{max} and G_{min} represent, respectively, the max and residual conductances while $V_{\frac{1}{2}}$ is the voltage at which 50% of conductance occurs.

For Na^+ inactivation curves the Boltzmann function becomes:

$$I/I_{max} = I_{min} + \frac{I_{max} - I_{min}}{1 + \exp^{\left(V_m - V_{\frac{1}{2}}\right)/K}} \tag{15}$$

where $V_{1/2}$ is now the midpoint of inactivation.

The slope factor (K), which is expressed in mV, measures the steepness of the curve and represents the relationship between the channel activation (or channel inactivation) and the membrane voltage. Different K slopes reflect a different dependence of the gating channel from the membrane potential (**Figure 3D**).

Finally, for K^+ activation curve and its IV relationship Eq. (15) can be written as:

$$I = I_{max} + \frac{I_{min} - I_{max}}{1 + \exp^{\left(V_m - V_{\frac{1}{2}}\right)/K}} \tag{16}$$

Fitting curves to exponential growth and decay functions of different order provides information about the τ of decay and rise phase of currents (**Figure 2A** and **C**). A single exponential function can well represent a proportional rate of growth/decay of our function

$$I = A \exp^{-t/\tau} + I_0 + kt \tag{17}$$

e.g. the initial rise phase of potassium currents. A two-phase model, *i.e.* double exponential, is used when our measure is the result of the sum of a fast and slow exponential rise/decay *e.g.* the fitting of Na^+ recovery from inactivation

$$I/I_{max} = A_1 \exp^{-t/\tau_1} + A_2 \exp^{-t/\tau_2} + A_3 \tag{18}$$

4. Conclusions

The neuronal information transfer is a complex and nonlinear process that strictly depends on the biophysical properties of membranes. The active and passive properties determine the rate at which the potential changes influencing both the integration of synaptic signals and the speed of action potential propagation. In this chapter we analyzed the main membrane parameters influencing the electrical signaling of the cell. The most salient aspects can be summarized as follows:

- The membrane capacitance affects the time course of electrical signals;
- The input resistance determines the amplitude of passive changes in membrane potential;

- The membrane resting potential influences the voltage-dependent membrane conductances at both synaptic and neuronal levels;
- The steepness of activating and inactivating kinetics represents a measure of the responsiveness of voltage-dependent channels with respect to the membrane voltage;
- The magnitude of the *window voltage* indicates the channel unavailability and therefore the level of synchronization of the channels.

Author details

Silvia Santillo
Istituto di Scienze Applicate e Sistemi Intelligenti "E. Caianiello", Pozzuoli, Italy

*Address all correspondence to: silvia.santillo@cnr.it

References

[1] Lindau M, Neher E. Patch-clamp techniques for time-resolved capacitance measurements in single cells. Pflügers Archiv. 1998;**411**:137-146. DOI: 10.1007/BF00582306

[2] Neher E, Sakmann B. The patch clamp technique. Scientific American. 1992;**266**(3):44-51. Available from: https://www.jstor.org/stable/24938980

[3] Molleman A. Patch Clamping: An Introductory Guide to Patch Clamp Electrophysiology. Wiley; 2003. 171 p. ISBN: 978-0-471-48685-5. Available from: https://onlinelibrary.wiley.com/doi/book/10.1002/0470856521

[4] Kornreich BG. The patch clamp technique: Principles and technical considerations. Journal of Veterinary Cardiology. 2007;**9**(1):25-37. Available from: https://pubmed.ncbi.nlm.nih.gov/17689466/

[5] Hill CL, Stephens GJ. An introduction to patch clamp recording. In: Dallas M, Bell D, editors. Patch Clamp Electrophysiology. Methods in Molecular Biology. Vol. 2188. New York, NY: Humana; 2021. DOI: 10.1007/978-1-0716-0818-0_1

[6] Kodirov SA. Whole-cell patch-clamp recording and parameters. Biophysical Reviews. 2023;**15**:257-288. DOI: 10.1007/s12551-023-01055-8

[7] Di Maio V, Ventriglia F, Santillo S. A model of cooperative effect of AMPA and NMDA receptors in glutamatergic synapses. Cognitive Neurodynamics. 2016;**10**(4):315-325. DOI: 10.1007/s11571-016-9383-3

[8] Di Maio V, Ventriglia F, Santillo S. AMPA/NMDA cooperativity and integration during a single synaptic event. Journal of Computational Neuroscience. 2016;**41**(2):127-142. DOI: 10.1007/s10827-016-0609-5

[9] Tsubo T. Analysis of the mechanism of synaptic integration focusing on the charge held in the spine. Biophysics and Physicobiology. 2021;**18**:290-304. DOI: 10.2142/biophysico.bppb-v18.036

[10] Hernández-Balaguera E, Vara H, Polo JL. Identification of capacitance distribution in neuronal membranes from a fractional-order electrical circuit and whole-cell patch-clamped cells. Journal of the Electrochemical Society. 2018;**165**(12):G3104-G3111, 165:12. Available from: https://iopscience.iop.org/article/10.1149/2.0161812jes

[11] Cole K, Curtis H. Electrical impedance of nerve during activity. Nature. 1938;**142**:209-210. DOI: 10.1038/142209b0

[12] Hodgkin AL, Huxley AF. A quantitative description of membrane current and its application to conduction and excitation in nerve. The Journal of Physiology. 1952;**117**(4):500-544. DOI: 10.1113/jphysiol.1952.sp004764

[13] Gentet LJ, Stuart GJ, Clements JD. Direct measurement of specific membrane capacitance in neurons. Biophysical Journal. 2000;**79**(1):314-320. DOI: 10.1016/S0006-3495(00)76293-X

[14] Di Maio V, Santillo S, Sorgente A, Vanacore P, Ventriglia F. Influence of active synaptic pools on the single synaptic event. Cognitive Neurodynamics. 2018;**12**(4):391-402. Available from: https://www.ncbi.nlm.nih.gov/pmc/articles/PMC6048015/

[15] Koch C, Rapp M, Segev I. A brief history of time (constants). Cerebral

Cortex. 1996;**6**(2):93-101. DOI: 10.1093/cercor/6.2.93

[16] Gillis KD. Techniques for membrane capacitance measurements. In: Sakmann B, Neher E, editors. Single-Channel Recording. Boston, MA: Springer; 1995. DOI: 10.1007/978-1-4419-1229-9_7

[17] Platzer D, Zorn-Pauly K. Letter to the editor: Accurate cell capacitance determination from a single voltage step: A reminder to avoid unnecessary pitfalls. American Journal of Physiology. Heart and Circulatory Physiology. 2016;**311**(4):H1072-H1073. Available from: https://api.semanticscholar.org/CorpusID:1484507

[18] Di Maio V, Santillo S, Ventriglia F. Multisynaptic cooperation shapes single glutamatergic synapse response. Brain Research. 2018;**1697**:93-104. Available from: https://pubmed.ncbi.nlm.nih.gov/29913131/

[19] Jahr C, Stevens C. Voltage dependence of NMDA-activated macroscopic conductances predicted by single-channel kinetics. The Journal of Neuroscience. 1990;**10**:3178-3182. DOI: 10.1523/JNEUROSCI.10-09-03178.1990

[20] Vargas-Caballero M, Robinson HP. Fast and slow voltage-dependent dynamics of magnesium block in the NMDA receptor: The asymmetric trapping block model. The Journal of Neuroscience. 2004;**24**(27):6171-6180. DOI: 10.1523/JNEUROSCI.1380-04.2004

[21] Mason MJ, Simpson AK, Mahaut-Smith MP, Robinson HP. The interpretation of current-clamp recordings in the cell-attached patch-clamp configuration. Biophysical Journal. 2005;**88**(1):739-750. DOI: 10.1529/biophysj.104.049866

[22] Perkins KL. Cell-attached voltage-clamp and current-clamp recording and stimulation techniques in brain slices. Journal of Neuroscience Methods. 2006;**154**(1-2):1-18. DOI: 10.1016/j.jneumeth.2006.02.010

[23] Aidley DJ. The Physiology of Excitable Cells. 3rd ed. Cambridge: Cambridge University Press; 1998. DOI: 10.1017/CBO9781139171182

[24] Di Resta C, Becchetti A. Introduction to ion channels. Advances in Experimental Medicine and Biology. 2010;**674**:9-21. Available from: https://link.springer.com/chapter/10.1007/978-1-4419-6066-5_2

[25] Vandenberg JI, Perry MD, Perrin MJ, Mann SA, Ke Y, Hill AP. hERG K+ channels: Structure, function, and clinical significance. Physiological Reviews. 2012;**92**(3):1393-1478. DOI: 10.1152/physrev.00036.2011

[26] Catterall WA. Structure and function of voltage-gated ion channels. Trends in Neurosciences. 1993;**16**(12):500-506. DOI: 10.1016/0166-2236(93)90193-P

[27] Keynes RD. The kinetics of voltage-gated ion channels. Quarterly Reviews of Biophysics. 1994;**27**(4):339-434. DOI: 10.1017/S0033583500003097

[28] Sigworth FJ. Voltage gating of ion channels. Quarterly Reviews of Biophysics. 1994;**27**(1):1-40. DOI: 10.1017/S0033583500002894

[29] Terlau H, Stühmer W. Structure and function of voltage-gated ion channels. Naturwissenschaften. 1998;**85**(9):437-444. DOI: 10.1017/S0033583500003097

[30] Tabarean IV, Juranka P, Morris CE. Membrane stretch affects gating modes of a skeletal muscle sodium channel. Biophysical Journal. 1999;**77**(2):758-774. DOI: 10.1016/S0006-3495(99)76930-4

[31] Tabarean IV, Morris CE. Membrane stretch accelerates activation and slow inactivation in Shaker channels with S3-S4 linker deletions. Biophysical Journal. 2002;**82**(6):2982-2994. DOI: 10.1016/S0006-3495(02)75639-7

[32] Jerng HH, Shahidullah M, Covarrubias M. Inactivation gating of Kv4 potassium channels: Molecular interactions involving the inner vestibule of the pore. The Journal of General Physiology. 1999;**113**(5):641-660. DOI: 10.1085/jgp.113.5.641

Chapter 4

Analyzing Dynamic Information from Single-Channel Recordings under Non-Stationary Conditions

Pedro Martin, Juan Ignacio Felice, Melisa Moncada, Nicolas Enrique and Veronica Milesi

Abstract

Ion channels are membrane proteins that facilitate the selective passage of ions, playing a key role in regulating cellular electrical properties and enabling a broad range of physiological functions. The patch-clamp technique has revolutionized our understanding of ion channel function by providing a detailed view of their activity. In particular, single-channel recordings have been fundamental in revealing the kinetic transitions between conductive and non-conductive states, which are essential for studying the biophysical, physiological, and pharmacological properties of ion channels. In physiological environments, ion channels are often exposed to dynamic conditions, such as fluctuating electrical signals and varying ligand concentrations. These non-stationary conditions require specialized analysis, which is critical for accurately describing the functional behavior of ion channels within their native cellular contexts. First latency (FL) analysis focuses on the time delay between a stimulus is applied and channel opening, providing information about channel activation kinetics. Open Dwell Time (ODT) analysis is particularly useful for studying the kinetics of inactivation, desensitization, and deactivation. Both offer valuable insights into the dynamics of channel transitions. This chapter presents several case studies demonstrating the application of FL and ODT analysis across different levels of channel complexity.

Keywords: ion channels, non-stationary analysis, first latency, open dwell time, patch-clamp

1. Introduction

Ion channels play a crucial role in cellular physiology by allowing the passive transport of ions across cell membranes. They are essential for various bioelectrical signals, including action potentials, receptor potentials, and postsynaptic potentials. Additionally, they are involved in synaptic processes, muscle contraction, secretion, and cell proliferation, among other cellular functions [1–3]. A canonical ion channel is a transmembrane protein that forms a pore with characteristic conductance and selectivity properties that can be highly restrictive, allowing only one type of ion to pass through, or less restrictive, as seen in non-selective cationic ion channels, conducting Na^+, K^+, and Ca^{2+}. Ion channels stochastically alternate between at least one closed conformational state (which prevents ion passage) and one open conformational state (which permits ion passage). Some voltage-operated and ligand-gated ion channels have an additional non-conductive state, called inactivated or desensitized, respectively. The average time a channel spends in each state can be modified by various stimuli, which include voltage, temperature, ligands, and mechanical stretching, depending on the channel's molecular structure and sensitivity to these factors.

Electrophysiological approaches have been fundamental to studying the properties of ion channels, such as the patch-clamp technique [4–6], a revolutionary tool that has dramatically enhanced our comprehension of ion channel function. This technique provides precise resolution of the complex dynamics of ion channel activity [1, 4, 7]. Patch clamping remains the only method capable of measuring, nearly in real time, ion channel activity in cell membranes. By employing various patch-clamp configurations [4], researchers can obtain macroscopic currents—through whole-cell, outside-out, or inside-out macropatch configurations—that reflect the activity of all ion channels present in the membrane. Alternatively, unitary currents can be recorded from cell-attached, inside-out, and outside-out configurations, allowing for the analysis of the behavior of individual protein molecules present in the patch membrane. Single-channel recordings enable the investigation of the biophysical properties of channels, such as ion selectivity, the rate of ion passage, and the duration of open, closed, inactivated, or desensitized states. In the case of voltage-activated ion channels, changes in cell membrane potential induce conformational rearrangement within the channel, resulting in a modified equilibrium between open and closed states. A conserved feature of these channels is the presence of a protein domain known as the voltage sensor domain (VSD), which is crucial for detecting electrical changes across the membrane. In canonical voltage-operated channels, the VSD is characterized by a series of positively charged amino acid residues located in the S4 transmembrane segment [8]. When a voltage change occurs within the channel's sensitivity range, it triggers a charge movement across the membrane that precedes the opening of the pore [1, 7, 9].

Furthermore, when a voltage stimulus is applied for an extended duration, allowing the channel to reach a steady-state condition, single-channel recordings can be used to determine the channel's open probability at that voltage, as well as the mean open time and mean closed time. However, immediately following the application of the stimulus and during the activation of the channel, there is a transitional phase where channels move from their initial condition (closed state) to the new steady-state condition. Understanding this transitional process requires a non-equilibrium analysis. This chapter aims to provide a comprehensive approach to studying ion channel behavior during the activation process at the single-channel level. In particular, following the application of a stimulus, we can measure the time elapsed until the first opening, denominated first latency (FL), which serves as an insightful parameter that characterizes the activation process of ion channels.

Inactivation, desensitization, and deactivation involve a transition from the open to a non-conductive state (inactivated, desensitized, and closed, respectively). These processes are essentially non-stationary. In these cases, since the transitions involve the departure from the open state, the kinetic behavior can be described by the time spent in this state before the channel closure, a parameter known as open dwell time (ODT).

In this chapter, we present and discuss the advantages of using FL analysis as a powerful tool for understanding and parameterizing the non-stationary activation process of ion channels. We will explore different cases with varying levels of complexity, demonstrating how FL analysis provides a precise and effective method for characterizing the activation dynamics of ion channels. Additionally, we will introduce ODT analysis as a means to describe voltage-dependent inactivation. This analytical approach can also be applied to other closing processes, such as deactivation and desensitization.

2. Why to use first latency to characterize the activation of an ion channel?

This section examines the simplest scenario for FL analysis: a cell expressing only the channel of interest (e.g., heterologously expressed in HEK and CHO cells, or *Xenopus laevis* oocytes) with minimal contamination from other channels. To represent this situation, in **Figure 1A**, we show the current obtained from a kinetic model where the channel presents three possible conformational states: two closed (C1 and C2) and one open state (O). In this hypothetical situation, the macroscopic current, is obtained by averaging multiple single-channel simulations. Thus, the first approach to obtain the kinetic parameters that characterize the activation process involves analyzing this macroscopic current. Particularly, in **Figure 1A**, the current changes during activation are consistent with a two-closed states model, inferred from fitting the current to a sum of two exponential functions using maximum likelihood estimation. This fitting generates two distinct time constants τ^{act} characterizing the transitions from C2 to O state (the faster) and from C1 to O (passing through C2, the slower). However, this procedure is biased since it does not take into account the following situation that may lead to distorted values. Observing the single-channel traces in **Figure 1A** (obtained with the same kinetic model), it is possible to observe that each individual channel needs a different time to activate. This is due to the stochastic nature of channel kinetics and the presence of different closed states. However, during the rising phase of the macroscopic current, there are channels that, after the stimuli, suffer a fast opening and return to the closed conformation and, even more, reopen previously to the time needed to reach the steady-state situation in the macroscopic current. Early opened channels that close and reopen before the other channels start to open, are included in the fitted activation of macroscopic current, and consequently, the value of the activation constant is distorted [10–12].

FL analysis allows us to overcome this limitation because it considers only the activation process of the channel and would result in a more precise study of the kinetics of channel activation.

3. How to calculate first latency?

To perform FL analysis, it is necessary to use single-channel recordings of the same channel stimulated several times under identical conditions. As we mentioned before, these recordings display variability in the time delay (latency) before the first channel opening occurs (see red lines in **Figure 1A**). This variability highlights the time-dependent nature of ion channel activation, which results from the asynchronous opening of channels. Additionally, this FL can be influenced by both the intensity of the stimulus and the initial state of the channels (C1 or C2). For instance,

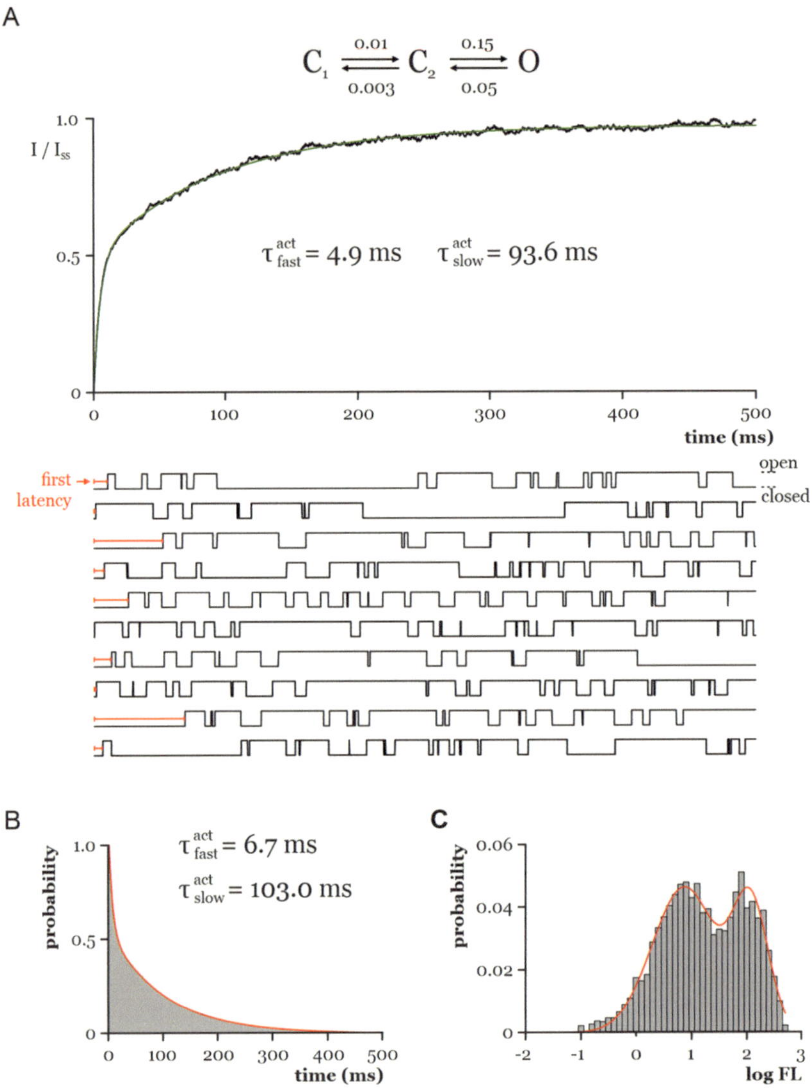

Figure 1.
First latency analysis in a simulated system, composed of only one type of channel with two closed and one open state. (A) Top panel: Kinetic model used for the simulations. Middle panel: macroscopic current obtained by averaging 4000 single-channel simulations, each lasting 500 ms. The superimposed green curve represents the fitted exponential function with two components. The time constants of these components are shown. Bottom panel: representative single-channel simulations with FL indicated in red. (B) Cumulative density function histogram of FL values, showing the probability of the channels remaining without openings as a function of the time elapsed after the stimulus application. The superimposed red curve represents the fitted exponential function with two components, and the time constants of the components are indicated. (C) Probability density function constructed from the logarithmic values of FL.

in voltage-gated ion channels, increased depolarizing stimulus magnitudes lead to shorter FL values.

The commercially available software allows obtaining the individual FL values from single-channel recordings, where the single-channel event detection tool, using the half-amplitude threshold criterion, is the most used for this aim. Once latencies are measured from each trace, we can obtain the activation time constant, considering only the activation process [11, 12]. To do this, we first construct a cumulative histogram from FL values, resulting in the cumulative density function (cdf, **Figure 1B**) [11–13]. The count represents the probability that the channel remains without openings at the different times recorded, as indicated in **Figure 1B**. Hence, in our example of a time-dependent voltage-channel activation, the probability of finding a closed channel (p) immediately after the beginning of the pulse (t = 0) is maximal (p = 1), but it decreases during the course of the voltage pulse, approaching p = 0. These distributions are then fitted to a sum of exponential decay functions as follows (Eq. (1)):

$$cdf = \sum_{i=1}^{n} A_i . e^{\frac{\mathrm{t}}{\tau_{\mathrm{i}}^{act}}} \tag{1}$$

In this expression, the number of components of the cdf (n) corresponds to the minimum number of detectable transitions through different closed states involved in the process of channel opening after application of the stimulus, while τ^{act} is the time constant of each component of channel activation. As expected for this example, our analysis of the cdf reveals two-time constant values corresponding to the channel transitions during the activation process, as the cdf could not be adequately fitted with only a single exponential function (**Figure 1B**). Moreover, to enhance the visualization and confirm the number of components indicated by the cdf, the FL histogram can also be constructed using the logarithmic values of the latencies, obtaining a probability density function (pdf). This latter representation effectively highlights the distinct components (see **Figure 1C**), and the parameters can be obtained by fitting the distribution with the sum of Gaussian functions. However, the constant values obtained from this representation are usually less precise than those obtained using the cdf, due to data transformation and the accuracy of the function used to fit the histograms.

Using the model in **Figure 1A**, we can calculate the real-time constants of the activation process when, before the stimulus, all the channels are in C1 (C1 to O transition, $\tau_{C1}^{act} = 104.5\, ms$) or in C2 (C2 to O transition, $\tau_{C2}^{act} =$ Ṭh ms). We observed that these time constants are very similar to those obtained from the cdf of FL values ($\tau_{slow}^{act} = 103.0\, ms$ and $\tau_{fast}^{act} = 6.7\, ms$) and all these values differ from the ones obtained from the macroscopic current ($\tau_{slow}^{act} = 93.6\, ms$ and $\tau_{fast}^{act} = 4.9\, ms$) which appears underestimated due to the phenomenon previously described. Despite this fact, it is important to note that in this case (a cell where our channel of interest is the only one present in the membrane), the macroscopic current approach provides a solution close to the real values and could be used as an acceptable approximation to parameterize the activation process. However, if the use of single-channel recording is mandatory (e.g., channels that are only visible in the cell-attached configuration) or when the construction of a kinetic model requires accurate constant values, then FL analysis is more adequate [11–13]. In the next sections, we will discuss other situations where the use of macroscopic currents to obtain the activation and inactivation time constants has more limitations and where FL analysis provides significant advantages for studying this non-stationary process.

4. Applying first latency analysis when multiple ion channels are present in the patch

FL analysis can also be applied to study the non-stationary properties of a particular ion channel in the context of a cell presenting other types of channels sensitive to the same stimuli. In this case, the macroscopic current represents the sum of the activation of all the different ion channels expressed in the cell. Thus, the time constants obtained from macroscopic current fitting do not represent the real kinetic process of our channel of interest. One way to obtain a more accurate activation parameter is by blocking or knocking the other ion channels that contaminate the macroscopic kinetics. However, these pharmacological or genetic interventions can also alter our channel of interest or even be impossible to perform. This is another situation where FL analysis can solve the problem inherent in macroscopic currents. Here, we present the FL analysis for the large-conductance voltage- and calcium-activated K^+ channel (BK or Slo1) present in human umbilical artery smooth muscle cells [14]. It will be used as an example to illustrate the method applied to a native cell.

In practice, to apply FL analysis in native cells, a depolarizing pulse must be applied consecutively at least 50–100 times [11, 14, 15]. The presence of the ion channel of interest is confirmed by its single-channel current amplitude at different voltages that reflects its own unitary conductance [14]. **Figure 2A** shows typical consecutively obtained inside-out single-channel recordings from a human umbilical artery smooth muscle cell where it is possible to observe the presence of different types of ion channels. In our experimental conditions, the larger current amplitude intensity corresponds to the BK channel (20 pA at +80 mV for a conductance of about 250 pS), and the other current amplitudes presumably correspond to a KV channel that has smaller conductance (see the #arrow in **Figure 2A**), and it is known to be present in this vascular smooth muscle cell [16].

Using these channel recordings, and as previously described, FL analysis involves measuring the elapsed time from the onset of the stimulus to the first opening of the BK channel, indicated by the red line in **Figure 2A**. This approach allows us to construct the cdf for the channel of interest, even in the presence of other channel types, since the cdf considers only the FL of the channel identified by its characteristic conductance. In this context, using the macroscopic current to obtain the kinetic activation constants is not appropriate, as it not only introduces the issues previously discussed but also aggregates activation events from other channels present in the cell membrane.

Moreover, in this realistic situation, it is possible to observe the presence of more than one copy of the channel of interest in our patch (indicated with the ¥arrow in **Figure 2A**). This fact should be considered at the moment of obtaining the time constant values because the probability of finding a channel opening in brief times increases with the number of channel copies in the patch. FL analysis allows the inclusion of recordings from membrane patches containing multiple copies of the channel of interest, but reliable cdf are only obtained when the membrane patch contains less than four or five channels. Otherwise, if the patch contains a larger number of channels, the counting of openings can lead to the omission of a longer exponential component of FL [11, 12]. Since the gating behavior of a single ion channel is independent of that of its neighbors, and the count values in cdf are expressed as probabilities, the real FL distribution (for membrane patches containing more than a single channel) can be obtained by taking the Nth root of the apparent distribution function, where N is the number of channels present in the patch [13]. The number of copies of the ion channel can be determined by a maximal stimulus to ensure multiple openings of all copies at the same time. In **Figure 2B**, we show the raw and corrected cdf obtained from a patch containing two BK channels. As expected, the calculated τ^{act} was longer (220 ms vs. 130 ms) when cdf was corrected.

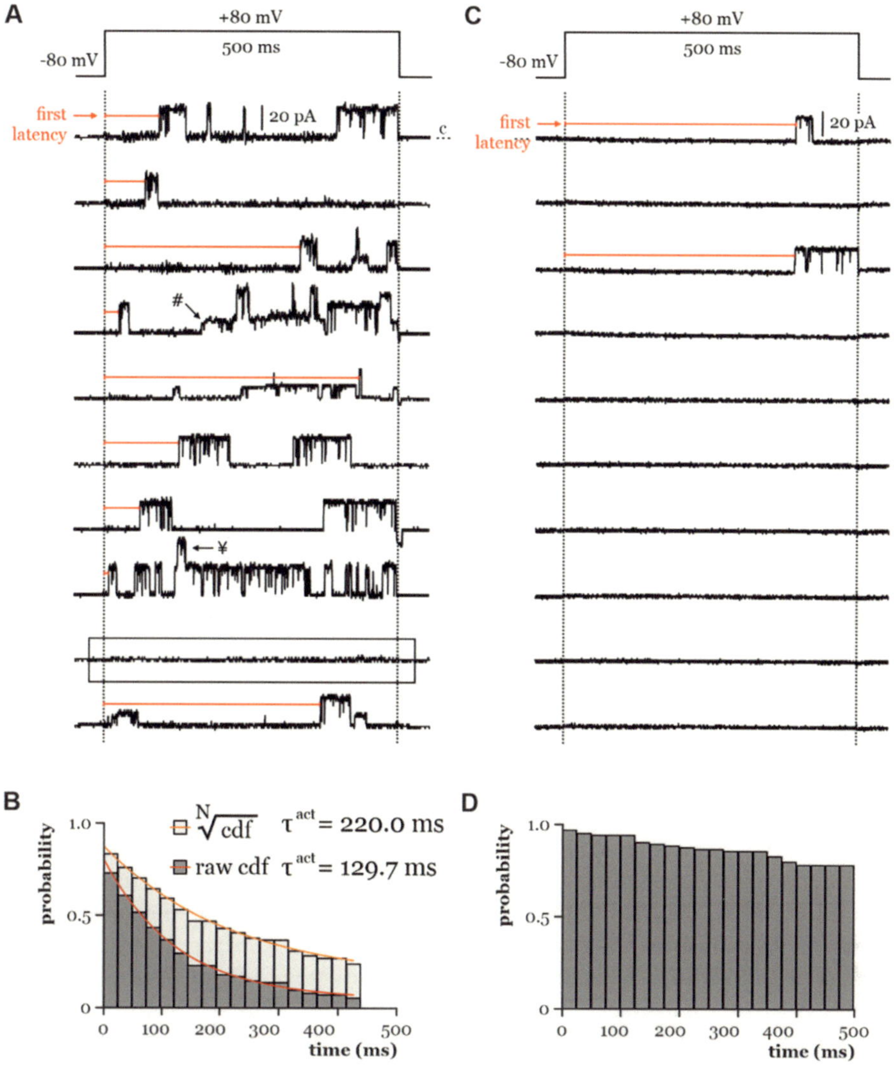

Figure 2.

First Latency analysis in native cells. (A) Representative single-channel recordings obtained in smooth muscle cells from human umbilical arteries by applying a 500 ms voltage step from a holding potential of −80 mV to a test potential of +80 mV in inside-out configuration under symmetrical K^+ conditions. "c" indicates the zero current level. The patch contained two copies (¥) of large-conductance voltage- and Ca^{2+}-activated K^+ channels (BK) identified by its amplitude value (¥ 22 pA), and a single copy of another ion channel with lower conductance (#). In each trace, the latency to the first BK channel opening is marked by a red line. The protocol was applied 100 times, and only in seven of these traces did the ion channel not open during the test pulse application (e.g., the second last trace). (B) Cumulative density function histogram of FL values, before and after correcting for the number of channels in the patch. The correction was achieved by applying the Nth root to the raw cumulative density function, where N is the number of BK channels in the patch. These distributions show the probabilities that the channel remained closed by the time indicated. Activation time constants of raw and corrected cdf were obtained by exponential fitting (red and orange lines, respectively). (C) Representative single-channel recordings under the same conditions as in panel A. The protocol was applied 100 times, and in 78 of these traces, the ion channel did not open during the activator pulse application. (D) Cumulative density function histogram of FL values corresponding to the patch shown in C. It was not possible to fit the histogram to obtain the time constants of the BK channel in this patch.

As we mentioned before, FL analysis requires experimental data where single-channel recordings are obtained through the repetitive application of the same stimulus to the patch. Moreover, a well-defined cdf is essential for identifying all the possible components involved in channel activation. Thus, stimulus intensity and duration must be carefully chosen to ensure at least one channel opening in most recordings. However, if the channel fails to open in only a few of the single-channel recordings, FL analysis can equally estimate the time constant of activation. In these cases, the FL value is considered longer than the pulse duration, and all traces are included in cdf construction. The presence of traces without channel openings will be evident at longer times of cdf, where the probability approaches zero but does not reach it (**Figure 2B**). Despite that, the time constant values can be obtained by the exponential fitting of the resulting cdf. Importantly, if a high proportion of traces without channel opening are included in the cdf, the results can potentially omit one of the exponential components of FL with a time constant longer than the pulse duration, or the obtained cdf could be impossible to be fitted with an exponential function (**Figure 2C** and **D**) [15]. Lengthening the stimulus duration and/or increasing the stimulus intensity to achieve a new state with a higher open probability are possible experimental strategies to ensure at least one channel opening in most of the recordings.

However, in some cases, it is impossible to obtain experimental conditions for the construction of the cdf. The intensity, duration, and repetition of the stimulus could be limited by the membrane seal stability. In this case, the number of individual FL values will be low, and histograms may not be well-defined. When this happens, we can calculate the mean FL value from the FL analysis as a good estimation of the non-stationary channel activation process. It is calculated as the average of the individual FL values and provides a useful parameter to analyze and compare different experimental conditions that could alter the ion channel activation kinetics. It is important to consider that using the mean FL value as a representative parameter of channel activation presents some limitations: (i) it can be calculated in patches containing only one copy of the channel of interest. If the patch presents more than one copy, the mean FL value will be underestimated. As we mentioned before, the probability of finding a channel opening at short times is higher when multiple channels are present in the patch. This fact is illustrated in **Figure 2A**, where the recording shows the activity of two BK channel copies. In this case, the calculated mean FL value was 119 ms, which is a shorter time compared to 220 ms, which is the real-time constant of activation obtained using the de cdf fitting. (ii) if traces without openings are not included, or are included with the pulse duration as FL value in both cases, the mean FL value will again be underestimated. This problem could be solved by modifying the recording conditions as we described before. (iii) The mean FL value is not suitable for the construction of kinetic models, since it tells us nothing about the number of transitions through different closed states involved in channel activation or about the time constants involved in these putative transitions.

5. Using first latency and open dwell time to characterize the kinetics of inactivating ion channels

In the case of voltage-dependent ion channels that present inactivation, single-channel recordings permit the independent study of the two non-stationary processes triggered by a depolarizing stimulus: channel activation followed by a fast inactivation, both with their own kinetic. FL analysis is similar to that used for non-inactivating ion channels described before and provides a precise constant value of channel opening kinetics. Simultaneously, the inactivation kinetics can be analyzed using the ODT obtained from the same recordings.

In this section, we present the analysis of the two aforementioned processes using a simulated voltage-operated sodium current (**Figure 3A**) derived from a kinetic model where the channel transits between the closed, open, and inactivated conformational states.

The macroscopic current from **Figure 3A** shows the voltage-activated inward current evoked by a depolarizing voltage step of 10 ms long. As shown, the current increases from zero to a negative peak current value and then decreases to zero in approximately 5 ms, indicating that the channels have reached a fully inactivated steady-state equilibrium. In this biphasic behavior, the channels undergo transitions from closed states to an open state (activation) and subsequently from the open state to the non-conductive inactivated state (inactivation). Therefore, initially, in the rising phase of the current, activation significantly outweighs inactivation. After the peak, inactivation becomes the predominant process and current diminishes.

Similarly to the analysis in *Section 2*, the activation and inactivation kinetics of the macroscopic currents can be determined by fitting the trace. As shown in **Figure 3A**, the entire current was accurately adjusted by the sum of two exponential functions. Each exponential component represents a different process: activation, in the rising phase of the current, characterized by a single time constant of 0.3 ms, and inactivation, characterized by a slower constant of 1.6 ms in the decreasing phase of the current. While the fitting assumes non-overlapping activation and inactivation, single-channel recordings in **Figure 3A** reveal that this is not always the case. Multiple channels can activate and inactivate within the macroscopic activation time, leading to the peak of the macroscopic current when the rates of channel opening and closing are equal. Consequently, adjusting macroscopic kinetic to the peak current excludes the influence of channels with delayed opening, resulting in an apparent faster process of activation [10–12]. Additionally, using macroscopic current decay to indicate inactivation is similarly biased, as the time constant kinetics are influenced by late channel openings.

FL analysis effectively eliminates distortions in the activation time constants. By fitting the cdf, it is possible to isolate the transition from the resting state to the open state, excluding the effects of channel inactivation. This provides a precise measurement free from the influence of the inactivation process. **Figure 3B** shows the cdf of the FL values and their adjustment with an exponential function. The precision of this method, in this case, is evident when comparing the time constant values obtained by fitting the macroscopic current (0.3 ms) with those obtained from the FL analysis (1.5 ms). This significant difference confirms that the value obtained from the macroscopic current fitting is faster due to overlap with early inactivated channels, making FL analysis a better approach for describing channel activation. Real activation time constants were also determined by exponential fitting on FL values obtained from separate traces corresponding to the two possible initial closed states (C1 and C2). Both activation time constants were found to be rapid and comparable, thus indistinguishable using FL ($\tau_{C1}^{act} = 1.72\,ms$ and $\tau_{C2}^{act} = 0.98\,ms$).

The inactivation process can also be studied using single-channel recordings by measuring the ODT preceding to channel closure due to inactivation. Similar to the approach used in FL analysis, we can measure the duration of ODT (**Figure 3A**, blue lines) and calculate the mean open time. Time constants of inactivation can

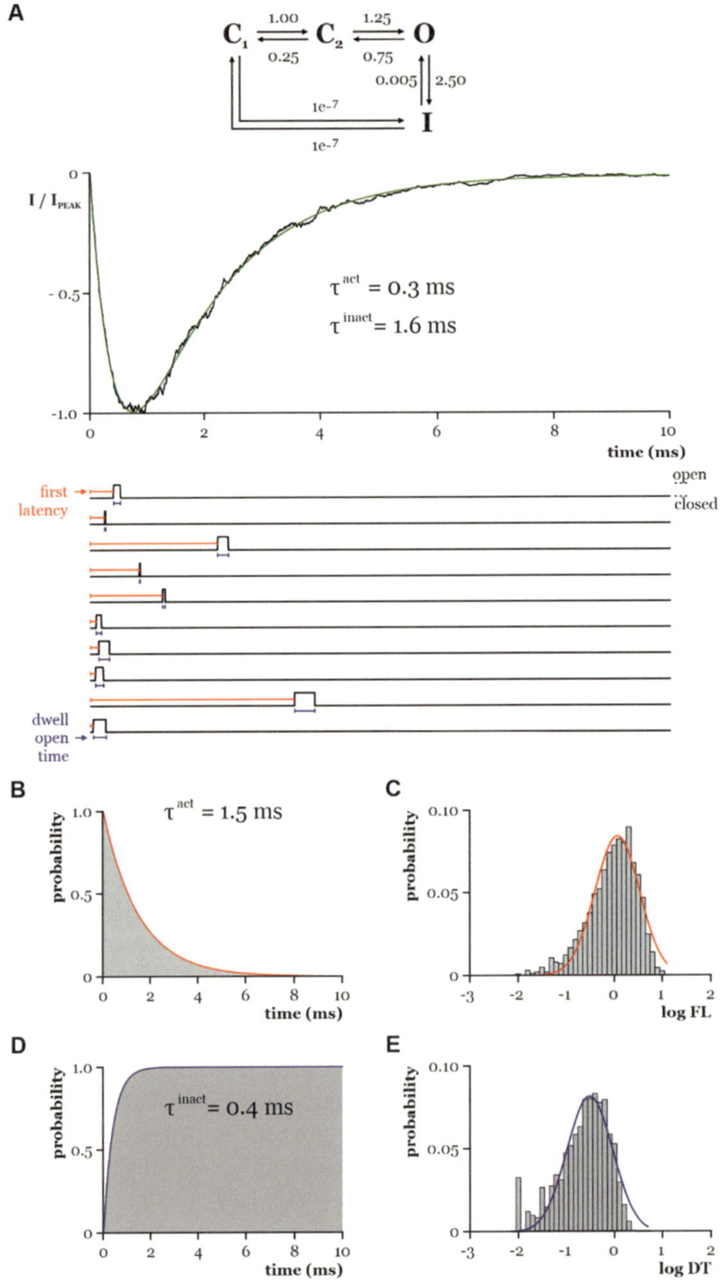

Figure 3.

First latency analysis in a simulated system, composed of only one type of channel with two closed, one open, and one inactivated state. (A) Top panel: Kinetic model used for simulations. Middle panel: macroscopic current obtained by averaging 4000 single-channel simulations, each lasting 10 ms. The superimposed green curve represents the fitted exponential function with two components (activation and inactivation). The time constants of these components are shown. Bottom panel: representative single-channel simulation with FL indicated in red and ODT in blue. (B) Cumulative density function histogram of FL values shows the probability of the channels remaining without openings as a function of time. The fitted exponential function is represented in red, and the time constant of activation is indicated. (C) Probability density function constructed from the logarithmic values of FL. (D) Cumulative density function histogram of ODT values. The fitted exponential function is represented in blue, and the time constant of inactivation is indicated. (E) Probability density function constructed from the logarithmic values of ODT.

be obtained from the cumulative density function of the measured ODT, as is shown in **Figure 3D** and **E** (for more details, see Chen et al. [15] and Sigworth et al. [17]). Single-channel analysis revealed a faster inactivation process than that inferred from macroscopic currents. The discrepancy in time constants (1.6 vs. 0.4 ms) highlights the impact of the late-opening channels on inactivation kinetics measurement.

Moreover, the kinetics of ligand-gated ion channels can be studied using the same methodology outlined in this chapter. Ligand binding induces channel activation through a conformational transition from the closed state to the open state, increasing the open probability. Afterward, channel activity rapidly declines to a desensitized, non-conducting state, even in the presence of the ligand. Although desensitization and inactivation differ in their mechanism, the experimental approach remains consistent with that used for inactivating voltage-gated ion channels.

Finally, ODT analysis is also applicable to other non-stationary processes, such as channel deactivation. In this process, the transition occurs from the open to the closed state upon stimulus removal. To measure ODT in this context, the recording conditions should ensure that the channel is opened when the stimulus ends. Then, the ODT is measured as the interval stimulus removal and the channel closure. Data manipulation and analysis for deactivation are similar to those described before, except that the time constant is referred to as the deactivation constant.

6. Conclusions

Analyzing ion channel behavior under non-stationary conditions is crucial for understanding their biophysical, physiological, and pharmacological properties. FL and ODT analysis are powerful tools for quantifying these transitional processes. Constructing a kinetic gating mechanism provides insights into the conformational changes underlying channel activation, deactivation, inactivation, and desensitization. In cell physiology, numerous and simultaneous stimuli are present at short time scales, where ion channel activation does not reach the steady-state condition; thus it is worthy to test ion channel activity in non-stationary condition. The action potential is a prime example, where the ion channel activity involved in the fast depolarization and repolarization is better represented by non-stationary analysis than stationary conditions. For instance, during the repolarization phase, a small fraction of potassium channels open, but this is sufficient to repolarize the cell. Likewise, many drugs such as antiepileptic and antiarrhythmic, can modify the shape and duration of action potentials due to changes in the ion channel activation kinetics in non-stationary conditions.

In conclusion, we underscore the importance of knowing that FL and ODT analysis are very useful in understanding the function of ion channels in complex biological processes.

Notes

Single-channel simulations were performed using the Gillespie algorithm [18], choosing arbitrary constant values to reproduce macroscopic currents that are consistent with experimental observations.

Author details

Pedro Martin[1], Juan Ignacio Felice[1], Melisa Moncada[2], Nicolas Enrique[1] and Veronica Milesi[1*]

1 Ion Channels Laboratory, Institute of Immunological and Physiopathological Studies (IIFP CONICET National University of La Plata), La Plata, Argentina

2 Center for the Study of Parasitology and Vectors (CEPAVE CONICET National University of La Plata), La Plata, Argentina

*Address all correspondence to: veronica@biol.unlp.edu.ar

References

[1] Hille B. Ion Channels of Excitable Membranes. 3rd ed. Oxford, New York: Oxford University Press; 2001. 816 p

[2] Hodgkin AL, Huxley AF. The components of membrane conductance in the giant axon of Loligo. The Journal of Physiology. 1952;**116**(4):473-496

[3] Nelson MT, Brayden JE. Regulation of arterial tone by calcium-dependent K^+ channels and ATP-sensitive K^+ channels. Cardiovascular Drugs and Therapy. 1993;**7**(Suppl 3):605-610

[4] Hamill OP, Marty A, Neher E, Sakmann B, Sigworth FJ. Improved patch-clamp techniques for high-resolution current recording from cells and cell-free membrane patches. Pflügers Archiv. 1981;**391**(2):85-100

[5] Neher E, Sakmann B, Steinbach JH. The extracellular patch clamp: A method for resolving currents through individual open channels in biological membranes. Pflügers Archiv. 1978;**375**(2):219-228

[6] Sakmann B, Neher E. Patch clamp techniques for studying ionic channels in excitable membranes. Annual Review of Physiology. 1984;**46**:455-472

[7] Sakmann B, Neher E, editors. Single-Channel Recording [Internet]. Boston, MA: Springer US; 1995 [cited 2024 Sep 10]. DOI: 10.1007/978-1-4419-1229-9

[8] Long SB, Campbell EB, Mackinnon R. Voltage sensor of Kv1.2: Structural basis of electromechanical coupling. Science. 2005;**309**(5736):903-908

[9] González C, Baez-Nieto D, Valencia I, Oyarzún I, Rojas P, Naranjo D, et al. K+ channels: Function-structural overview. Comprehensive Physiology. 2012;**2**(July):2087-2149

[10] Hille B. Gatting: Voltage sensing and inactivation. In: Ion Channels of Excitable Membranes. 3rd ed. Oxford, New York: Oxford University Press; 2001

[11] Aldrich RW, Corey DP, Stevens CF. A reinterpretation of mammalian sodium channel gating based on single channel recording. Nature. 1983;**306**(5942):436-441

[12] Aldrich RW, Yellen G. Analysis of nonstationary channel kinetics. In: Sakmann B, Neher E, editors. Single-Channel Recording [Internet]. Boston, MA: Springer US; 1983. pp. 287-299 [cited 2024 Sep 10]. DOI: 10.1007/978-1-4615-7858-1_13

[13] Colquhoun D, Hawkes AG. The principles of the stochastic interpretation of ion-channel mechanisms. In: Sakmann B, Neher E, editors. Single-Channel Recording [Internet]. Boston, MA: Springer US; 1983. pp. 135-175 [cited 2024 Sep 10]. DOI: 10.1007/978-1-4615-7858-1_9

[14] Martín P, Moncada M, Enrique N, Asuaje A, Valdez Capuccino JM, Gonzalez C, et al. Arachidonic acid activation of BKCa (Slo1) channels associated to the β1-subunit in human vascular smooth muscle cells. Pflügers Archiv—European Journal of Physiology. 2014;**466**(9):1779-1792

[15] Chen RS, Geng Y, Magleby KL. $Mg^{(2+)}$ binding to open and closed states can activate BK channels provided that the voltage sensors are elevated. The Journal of General Physiology. 2011;**138**(6):593-607

[16] Martín P, Rebolledo A, Palomo ARR, Moncada M, Piccinini L, Milesi V. Diversity of potassium channels in human umbilical artery smooth muscle cells: A review of their roles in human umbilical artery contraction. Reproductive Sciences. 2014;**21**(4):432-441

[17] Sigworth FJ, Sine SM. Data transformations for improved display and fitting of single-channel dwell time histograms. Biophysical Journal. 1987;**52**(6):1047-1054

[18] Gillespie DT. Exact stochastic simulation of coupled chemical reactions. The Journal of Physical Chemistry. 1977;**81**(25):2340-2361

Chapter 5

Patch Clamp and Ion Channel Dynamics

Amina Asghar Chaudhry, Sumaira Kanwal and Shazia Perveen

Abstract

The automated patch-clamp approach allows researchers to directly examine drug effects on ion channels but cannot forecast sodium channel inhibitors. Created by Erwin Neher and Bert Sakmann in the 1970s, it enabled the first recording of single ion channel molecule currents increasing our understanding of channel function. The study aimed to determine the current contributions of separate ionic channels in a biological membrane using heat-polished micropipettes placed on enzyme-washed cells. The pipettes were filled with saline and had larger apertures than standard microelectrodes. The main challenge was creating an electrical seal between the membrane and pipette. The discovery that using clean pipettes and applying suction can increase the pipette membrane seal by 100–1000 is a significant advancement in this area. The membrane patch is drawn into the pipette's interior creating an omega-shaped semi-vesicle that adheres securely to the inner pipette wall. This allows for high-resolution measurements in even the smallest cells including red blood cells. The study also explored the use of patch pipettes for studying neurotransmitters or drugs with the cell-attach patch being the most suitable method. Whole-cell patch-clamp recordings were used for detailed observations of rapid conductance alterations induced by ion channels across a neuronal membrane. Temperature, pressure, pH, ion concentration and intracellular signaling can all affect the dynamics.

Keywords: currents, seal, channels, clamp, membrane patch, ion, resistance

1. Introduction

The automated patch-clamp approach has made it possible to directly examine the effects of different drugs on ion channels. But when it comes to sodium channel inhibitors, IC50 values or even KR and KI values cannot be used to forecast the therapeutic potential of any particular molecule [1].

1.1 Patch clamp

Ion channels are expressed by all cells. The most common applications of patch-clamp methods include the study of neurons, muscle fibers, cardiomyocytes and oocytes that overexpress single ion channels. The voltage clamp method can be used for patch clamping. In this instance, the researcher controls the voltage across the cell membrane and records the currents that arise. As an alternative, one can employ the current clamp method. Here, the researcher controls the passage of current across the membrane and the resulting voltage variations are recorded usually as action potentials [2].

1.2 History

The patch clamp was created by Erwin Neher (German biophysicist with expertise in cell physiology) and Bert Sakmann (German cell physiologist) late in the 1970s and early in the 1980s. With the first recording of single ion channel molecule currents made possible by this discovery, our knowledge of the role that channels play in basic cell functions including action potentials and nerve activity has increased. For their achievement, Neher and Sakmann were awarded the 1991 Nobel Prize in Physiology or Medicine [3].

2. The fundamental idea and variations

Resolving the separate ionic channel current contributions in a biological membrane was the first objective. In order to do this, heat-polished micropipettes were placed on the surface of enzyme-washed cells and the current passing through the pipette was measured. The pipettes were filled with regular saline unlike microelectrodes and had apertures that were somewhat bigger than those of standard microelectrodes (0.5–1 μm). The primary technological challenge included creating an electrical seal between the membrane and pipette. Furthermore, the main source of noise was the shunt conductance between the water and the pipette interior. Large leakage currents across the shunt have to be prevented at all times by holding the pipette at the bath potential.

An important development in this field is the finding that the pipette membrane seal may be increased by a factor of 100 to 1000 by utilizing clean pipettes and delivering a tiny bit of suction on the pipette upon contact. As a result, a little membrane patch is drawn into the pipette's interior creating an omega-shaped semivesicle. This adheres securely to the inner pipette wall in an all-or-nothing manner. At that point, the membrane is securely fastened to the pipette walls and the electrical resistance that exists between 10 and 100 GΩ across the seal. One may apply voltages across the membrane (and the seal) and measure currents with high resolution by making use of the high resistance of the seal. A limited amount of mechanical manipulations may be carried out because of the solid attachment [4].

There are three simple mechanical interventions that may be performed raising the pipette momentarily out of solution rupturing the membrane patch with a brief suction pulse, and withdrawing the pipette. The pipette membrane seal holds for all three operations for a sufficient number of trials resulting in clearly defined configurations. Breaking the membrane patch creates an electrical connection between the pipette and the inside of the cell. The pipette-cell assembly and the bath are well insulated. This approach, which we call "whole-cell measurement", looks like a more delicate version of a typical microelectrode penetration and yields high-quality recordings in even the smallest cells including red blood cells. Use pipettes filled with high-K+ and low-Ca2+ fluid if entering the cell is the aim.

When the pipette is removed from the cell-attached form, it often reaches an intermediate state when the pipette tip becomes sealed with a closed vesicle.

A membrane-enclosed cytoplasmic bridge is initially removed from the cell during withdrawal. This narrows and eventually collapses as the pipette and cell get farther apart leaving the pipette with a closed structure inside and the cell intact. Electrical measurement is not appropriate for this vesicle [5].

Short air exposure, however, will shatter its outer layer creating an "inside-out" patch. The reason for choosing this term is that the membrane's previous cytoplasmic surface is now facing the bath solution. If the withdrawal is carried out in conditions free of calcium inside-out patches also referred to as "excised patches" are readily obtained without being exposed to the air. An "outside-out" patch made with an identical configuration after the pipette was taken out of a "whole-cell" setup. While changing the bath solution is typically straightforward changing the pipette solution during an experiment necessitates thoughtful preparation. Here are where outside-out and inside-out patches work best together [6].

2.1 The cell-attach patch

In order to achieve, using this process, a gigaseal sealing with resistance to electricity on an order of a gigaohm pipette is sealed onto the cell membrane while making sure the membrane stays intact. This makes it possible to record currents flowing through one or more of the ion channels in the membrane patch that the pipette was able to capture. There is relatively minimal disruption of the cell structure since it just attaches to the cell membrane's outside. Furthermore, any intracellular processes that typically influence the channel will continue to work as they would physiologically provide that the interior of the cell is not disrupted. Finding the ideal configuration with this approach is likewise rather simple and once it is, it is reasonably stable.

When studying neurotransmitters or drugs that are under investigation the pipette solution containing ligand-gated ion channels or channels controlled by metabotropic receptors is frequently supplemented. Here, it can interact with the membrane that was formerly the membrane's outside. The medication being used can be identified by the channel activity that results [7]. However, it is typically not feasible to alter the drug concentration within the pipette at that point. This means that the approach is only applicable to a single patch's dosage response curve point. Therefore many cells and patches are used to achieve the dosage response. On the other hand, voltage-gated ion channels are repeatedly clamped in a single patch at various membrane potentials. The fact that the cell's internal pathways cannot be directly altered while being undisturbed is another possible disadvantage of this method.

2.2 Whole cell recording

Detailed observations using whole-cell patch-clamp recordings of the rapid conductance alterations are induced by ion channels across a neuronal membrane. Glass micropipettes with a significantly bigger aperture at the tip are utilized as patch pipettes in place of sharp micropipettes for recording entire cells. As a result, the electrode resistance of saline is often a few megaohms which is at least an order of magnitude lesser than that of sharp electrodes. The patch electrode is likewise filled with an interior fluid as with sharp intracellular electrodes. *In vitro* investigations use infrared video microscopy to make patch recordings on brain slices or cultured neurons while following visual instructions.

An infrared light source illuminates the sample and an infrared camera visualizes it, which has an amplifier for current and voltage clamp recordings and micromanipulators for electrode placement [8]. A micromanipulator transfers the patch pipette to the cell of interest to record or "patch" it. Visual training is made possible by the patch pipette's increased tip size.

2.3 Out-side out patch clamp

The optimal patch configuration for examining ionic channels regulated by externally placed receptors is the outside-out patch arrangement. It allows for the confirmation of the patch's orientation. As in a conventional whole-cell arrangement, the development of outside-out membrane patches begins with the creation of a gigaohm seal. Next, the membrane underneath the patch is ruptured using negative pressure and finally the patch pipette is progressively removed from the cell as the membrane splits [9]. The membrane will then fold onto itself and close resulting in the development of an excised outside-out membrane patch. Sometimes, the membrane will fold incorrectly resulting in an inside-out membrane patch. Given that the patch solution in this configuration is the intracellular solution, it must contain the ATP, GTP and cAMP intercellular components that may be required for the appropriate functioning of the ion channel or signal mechanism under study.

2.4 Inside-out patch clamp

This patch configuration is commonly used to study extracellular ligand-gated ion channels. The bath solution may be used to introduce ligands of this kind to the patch. To create an inside-out membrane patch, first create a gigaohm seal. Next, until the membrane splits remove the patch pipette from the cell gently. A vesicle is then formed when the membrane folds in on itself. The membrane vesicle's exterior face is then burst by either briefly coming into touch with a paraffin droplet or dried Sylgard exposed to low Ca2+ solution or quickly passing it across the bath solution-air interface [10]. The extracellular solution is often present in the patch-pipette when employing an inside-out patch configuration. If the patch pipette includes agonists for extracellular ligand-gated ion channels this recording device can be utilized to research these channels. Since the chamber's cytoplasmic membrane face is immersed in the solution, Ca2+, spermidine and cGMP are a few examples of intracellular modulators or activators of the ion channels in the patch that may be investigated by bath administration. This arrangement is perfect for researching channels that intracellular signals activate or alter. It is important to remember that in this setup the patch pipette potential and the membrane potential are equal and opposite to each other. Furthermore, inward currents have characteristics comparable to outward currents such as the cell-attached patch [11].

3. Factors affecting ion channel dynamics

Indeed a wide range of parameters including temperature, pressure, ion concentration, pH and intracellular signaling affect the dynamics of route clumping and ion channels. All of these variables have the power to modify the ion channel conformational states and activities which can alter cellular excitability and signaling [12].

- *Temperature:* Elevated temperatures have the ability to augment ion channel activity by augmenting the kinetic energy of molecules. On the other hand, high heat can cause denaturation.
- *Pressure:* Ion channel permeability and gating processes can be impacted by conformational changes brought about by pressure changes.
- *Ion Concentration*: The driving force for ion flow and the opening/closing of channels can be directly impacted by the concentration of certain ions such as Na+, K+ or Ca2+.
- *pH*: Changes in pH can influence the protonation states of residues inside or close to ion channels which can alter their activity.
- *Intracellular Signaling*: Channel dynamics can be changed by phosphorylation or other alterations that result from pathways involving second messengers such as cAMP or Ca2+ [13].

4. Visualization techniques

Different visualization approaches can be used to examine these dynamics.

4.1 Fluorescence microscopy

One effective technique for seeing ion channel kinetics and cellular activities in real time is fluorescence microscopy. Here are some important things to think about.

Fluorescent proteins like GFP or certain fluorescent dyes that attach to them are used to tag ion channels. This makes the channels visible to researchers using a fluorescent microscope. The fluorescent tags are excited by the microscope using particular light wavelengths causing them to release light with a longer wavelength. Images are made by capturing this emitted light. Researchers can measure how rapidly fluorescently tagged ion channels open and close in response to stimuli by observing the mobility and behavior of these channels. The distribution of ion channels throughout the membrane and their kinetics during cellular signaling events are ascertained with the use of fluorescence microscopy. In order to help in drug development, this approach is used to investigate how different pharmacological agents alter ion channel dynamics [14].

4.1.1 Taking into account

- *Photostability:* Selecting fluorescent tags that are more photostable than others is important as it might impact the data quality.
- *Live-cell imaging:* Because ion channel behavior can vary greatly depending on the circumstances, it is crucial to preserve cell viability during the imaging process.
- *Data analysis:* To understand the dynamics and interactions shown advanced imaging calls for sophisticated data analysis methods.

4.2 Electrophysiology

Patch-clamp recordings and other techniques allow for comprehensive measurements of ion currents shed light on gating and channel dynamics, being aware of neuronal excitability, synaptic transmission and the electrical properties of brain networks. The electrical activity of the heart including cardiomyocyte action potentials and arrhythmias is examined. Assessing how medications affect cellular excitability and ion channel activity is essential for creating therapeutic interventions for a range of illnesses. Examining the electrical characteristics of muscle cells particularly the mechanism by which action potentials initiate contraction. Accurate measurements depend on maintaining cell viability calls for cautious handling and ideal environmental conditions, in order to ensure precise data gathering and analysis. Electrophysiological methods require specialized equipment and knowledgeable staff. Understanding channel kinetics and cell behavior is often necessary for the interpretation of electrical signals calls for in-depth investigation [15].

4.3 Super-resolution microscopy

Individual ion channels may be seen with great spatial resolution using methods like STED or PALM can also show the distribution and interactions between them. High spatial resolution imaging of ion channels is made possible by this technique. STED can pinpoint the exact location of ion channels in the membrane by reducing fluorescence surrounding the focus point. Photo activatable fluorescent proteins may be utilized by PALM to tag ion channels allowing researchers to monitor individual channels throughout time to recreate their dynamics and mobility. STORM facilitates high-resolution ion channel visualization much like PALM. It takes pictures of individual molecules using a stochastic technique revealing information about their relationships and distribution. In a more live-cell-friendly way, SIM can yield useful information regarding the kinetics and organization of ion channels even if its resolution is not as high-resolution as that of STED or PALM/STORM [16]. The precise positions of ion channels inside the membrane may be mapped using super-resolution methods that can show organization and clustering that may have an impact on function. Researchers can learn more about the dynamics of ion channels such as how they move in response to stimuli or during cellular processes by monitoring their movement over time. Certain super-resolution methods can provide light on the conformational configurations of ion channels hence aiding in the understanding of their functional states and gating processes. Studying the interactions between ion channels and other proteins using FRET (Fluorescence Resonance Energy Transfer) and super-resolution microscopy might provide light on signaling cascades and biological reactions. Although numerous super-resolution methods are applicable to fixed samples phototoxicity and the requirement for certain conditions to sustain cell viability make it difficult to get relevant data from living cells. Making the right fluorescent tag selections is essential to reducing background noise and increasing signal-to-noise ratios. Super-resolution data analysis can be challenging requiring sophisticated software and image processing knowledge to reliably decipher dynamics and interactions [17].

4.4 Automated high-throughput screening

Large-scale data gathering is made possible by automated systems' ability to quickly test a variety of situations (such as changing temperature or ion concentrations) and record the resulting ion channel behavior. With the use of HTS, hundreds of compounds may be screened to find those that alter ion channel function. This process might result in the creation of novel treatments for diseases including cardiac arrhythmias, pain and epilepsy [6]. Comprehensive analysis of the effects of different substances on the kinetics and gating characteristics of ion channels is made possible by automated technologies. Those that use phenotypic screening to discover new modulators and pathways can be aided by screening for substances that cause certain physiological reactions connected to ion channel function. To comprehend ion channel dynamics within the context of particular illnesses, high-throughput techniques might be applied to patient-derived cells or disease models [18].

4.5 Computational modeling

In addition to actual data, simulations can offer predictions on ion channel behavior under different circumstances. The interaction between putative pharmacological candidates and ion channels may be predicted by computational models help find efficient modulators. By simulating how mutations affect ion channel function models can shed light on hereditary channelopathies and other illnesses. The outputs of simulations can assist forecast outcomes and optimize circumstances for electrophysiological research which can then be used to influence experimental design. The effects of different ligands or changes on ion channel gating and dynamics may be investigated using computational methods [19]. Significant processing power and time are needed for high-resolution simulations particularly MD which might be a constraint. The quality of the force fields and simulation parameters is a determining factor in accurate modeling. Careful validation against experimental data is necessary for this. It is difficult to simulate the intricacy of the complex biological settings in which ion channels function. The enormous volume of data produced by simulations calls for advanced analytic methods in order to derive biologically significant findings [20].

5. Application

- Used to investigate a cardio-specific blockage of the ATP-sensitive potassium channel in isolated ventricular myocytes from Guinea pigs.

- Used in Kidney cells.

- For the evaluation of antiarrhythmic agents.

- To recognize various kinds of calcium channels.

- To assess how well potassium channel openers work.

- Applicable to molecular biology.

- Studies using voltage clamps on $Na+$ channels

- Employed to explore an extensive array of electrophysiological cell characteristics.

6. Conclusions

An effective, heavily modified approach, this technology is being developed to help the drug development process with new methods that yield more precise and efficient information. This technology is being developed for fresh techniques to produce more effective and accurate information that supports the drug development process. It is sufficient to observe certain channel proteins on metabolically intact cells opening and closing repeatedly in real time. As of right now, no other technique provides a more precise or comprehensive knowledge of how calcium channel gating functions and is controlled.

Acknowledgements

All authors equally contributed in write up of this book chapter.

Conflict of interest

There is no conflict of interest.

Author details

Amina Asghar Chaudhry[1], Sumaira Kanwal[2] and Shazia Perveen[1]*

1 Women University Multan, Multan, Pakistan

2 COMSATS University, Sahiwal Campus, Islamabad, Pakistan

*Address all correspondence to: drshazia.zool@wum.edu.pk

References

[1] Hill CL, Stephens GJ. An introduction to patch clamp recording. In: Patch Clamp Electrophysiology: Methods and Protocols. 2021. pp. 1-19

[2] Bednarczyk P, Kampa RP, Gałecka S, Sęk A, Walewska A, Koprowski P. Patch-clamp recording of the activity of ion channels in the inner mitochondrial membrane. In: Mitochondrial Medicine: Volume 2: Assessing Mitochondria. New York, NY: Springer US; 2021. pp. 235-248

[3] Verkhratsky A, Parpura V. History of electrophysiology and the patch clamp. Patch-Clamp Methods and Protocols. 2014;**1183**:1-19

[4] Kumari A, Nguyen DM, Garg V. Patch-clamp technique to study mitochondrial membrane biophysics. The Journal of General Physiology. 2023;**155**:8

[5] Bell DC, Fermini B. Use of automated patch clamp in cardiac safety assessment: Past, present and future perspectives. Journal of Pharmacological and Toxicological Methods. 2021;**110**:107072

[6] Toh MF et al. Application of high-throughput automated patch-clamp electrophysiology to study voltage-gated ion channel function in primary cortical cultures. SLAS Discovery: Advancing the Science of Drug Discovery. 2020;**25**(5):447-457

[7] Kodirov SA. Whole-cell patch-clamp recording and parameters. Biophysical Reviews. 2023;**15**(2):257-288

[8] Gezelle J, Saggu G, Desai SA. Promises and pitfalls of parasite patch-clamp. Trends in Parasitology. 2021;**37**(5):414-429

[9] Obergrussberger A et al. Automated patch clamp in drug discovery: Major breakthroughs and innovation in the last decade. Expert Opinion on Drug Discovery. 2021;**16**(1):1-5

[10] Polder HR et al. Voltage-clamp and patch-clamp techniques. In: Practical Methods in Cardiovascular Research. Berlin, Heidelberg: Springer; 2005. pp. 272-323

[11] Brown APY, Greenberg HZE. Patch clamp. British Journal of Hospital Medicine. 2016;**77**(5):C74-C77

[12] András V et al. Cardiac transmembrane ion channels and action potentials: Cellular physiology and arrhythmogenic behavior. Physiological Reviews. 2021;**101**(3)

[13] Liebovitch LS, Krekora P. The physical basis of ion channel kinetics: The importance of dynamics. IMA Volumes in Mathematics and Its Applications. 2002;**129**:27-52

[14] Harms GS et al. Probing conformational changes of gramicidin ion channels by single-molecule patch-clamp fluorescence microscopy. Biophysical Journal. 2003;**85**(3):1826-1838

[15] McNeill J et al. Ion channels and electrophysiological properties of astrocytes: Implications for emergent stimulation technologies. Frontiers in Cellular Neuroscience. 2021;**15**:644126

[16] Bastian C et al. Super-resolution microscopy for the study of store-operated calcium entry. Cell Calcium. 2022;**104**:102595

[17] Navikas V et al. Correlative 3D microscopy of single cells using

super-resolution and scanning ion-conductance microscopy. Nature Communications. 2021;**12**(1):4565

[18] Bokhari FF, Albukhari A. Design and implementation of high throughput screening assays for drug discoveries. High-Throughput Screening for Drug Discovery. 2021;**3**:53

[19] Shine JM et al. Computational models link cellular mechanisms of neuromodulation to large-scale neural dynamics. Nature Neuroscience. 2021;**24**(6):765-776

[20] Heijman J et al. Computational models of atrial fibrillation: Achievements, challenges, and perspectives for improving clinical care. Cardiovascular Research. 2021;**117**(7):1682-1699

Chapter 6

Patch Clamping in Combination with Fluorescence Is a Reliable Technique for Studying Pore-Forming Ion Channels

Guilherme Pegas Teixeira, Leandro Rocha and Robson Xavier Faria

Abstract

Ion channels play essential roles in the functional balance of cells, whether in the context of electrical balance or in the activation of signaling pathways for specific processes. Pore formation is a notable feature of several channels, such as P2X7R, TRPV1, TRPA1, connexins, pannexin−1, and Maxi-anion, which significantly increase the permeability of the plasma membrane, allowing high ionic conductance and molecular weights of up to 900–1000 Da. Exploring the biophysical and physicochemical properties of these receptors is an important way to better understand their physiological and pathological functions. To this end, the patch clamp electrophysiology technique and methods using cationic and anionic fluorescent dyes are used to study these channels, making it possible to estimate and quantify pore size, ionic conductance, and selectivity. Therefore, in this work, we address different pore-forming channels through electrophysiological and fluorescence investigations of the physical and chemical properties of these large conductance channels.

Keywords: ion channels, fluorescent dyes, patch clamp, pore size, conductance

1. Introduction

The plasma membrane is a fundamental and complex structure for cellular life. Owing to their biophysical properties, specific solutes require facilitators to cross through the membrane and reach the intracellular environment [1]. In this sense, ion channels regulate the homeostasis of cations, such as Na^+, K^+, and Ca^{2+}, and anions, such as Cl^- and glutamate. Furthermore, they participate in the action potential by responding to changes in membrane voltage [2].

Pore-forming channels allow the passage of high molecules with positive and negative mono-, di-, tri-, tetra-, and pentavalent charges, with different physical characteristics according to the concentration and the electrochemical gradient. This phenomenon increases plasma membrane permeability for solutes of up to 1000 Da. For over 40 years, electrophysiology (patch clamp) and fluorescent dye uptake have been two techniques used to study these receptors [3–5].

Patch clamp is a versatile and reliable electrophysiological technique based on electrical recording of the total or a portion of the plasma membrane containing ion channels, thus measuring the ionic and electrical functionality of these receptors [6]. Dye uptake assays are most often used to evaluate channel pore size as conventional fluorescent dyes are charged and too large to pass through the membrane [3]. With both techniques, it is possible to estimate the "cutoff" of the studied receptor for the molecular weight in Dalton (Da) or the molecular diameter in Angstrom (Å). Some examples of large organic solutes and fluorescent dyes discussed throughout this work are shown in **Tables A1 and A2**. Therefore, we address the main pore-forming channels and their biophysical characteristics.

2. Pore-forming channels and their biophysical and molecular properties

As mentioned previously, patch clamp techniques combined with fluorescence methods are used to refine analyses of different targets. In the context of pore-forming channels, calcium homeostasis modulator 1 (CALHM1), connexins (Cx), pannexin-1 (Panx-1), the Maxi-anion channel (MAC), purinergic receptors (P2X2R, P2X4R, and P2X7R), plasma voltage-dependent anion-selective channel 1 (pl-VDAC1), and transient receptor potential (TRP channels: TRPV1, TRPV2, TRPV3, TRPV4, TRPA1, and TRPM8) are highlighted for their biophysical properties via patch clamp recordings of whole-cell, cell-attached, inside-out, and outside-out signals, in addition to fluorescent dye uptake. Information about the biophysical data from these channels can be found in **Table 1**.

2.1 Calcium homeostasis modulator 1 (CALHM1)

CALHM1, formerly FAM26C, is a Ca^{2+} channel belonging to the family of calcium homeostasis modulators. This receptor is associated with the control of Ca^{2+} in the central nervous system. These data point to the vital role of CALHM1 in the pathogenesis of Alzheimer's disease [23, 24]. CALHM1 induces cationic currents in cells and increases the cytoplasmic Ca^{2+} concentration in response to the removal of extracellular Ca^{2+}. Significant reductions in extracellular Ca^{2+} and CALHM1 open and strongly hyperpolarize the membrane voltage [9].

The cations Na^+ and K^+ and the anion Cl^- also pass through CALHM1 in the following order of selectivity: $Ca^{2+} >> Na^+ \sim K^+ > Cl^-$. In the case of Na^+, the relative permeability ratio concerning Ca^{2+} and $P_{Ca}{}^{2+}$: PNa^+ is ~11 [25]. Whole-cell recordings in oocytes expressing CALHM1 were $P_{Na}{}^+/P_K{}^+/P_{Cl}{}^-$ 1:1.14:0.52 and 1:1.46:0.88 with or without 2 mM extracellular Ca^{2+}, respectively, suggesting that CALHM1 has a greater tendency for monovalent cations than anions but is weak for divalent cations other than Ca^{2+}, such as Mg^{2+} [9].

The protein structure of CALHM1 is similar to that of connexin (Cx), forming a hexamer and its monomers with four transmembrane domains. Although the structure is similar, CALHM1 cannot form GAP junctions. Interestingly, the permeability of the large organic monovalent cations tetramethylammonium (TMA^+), tetraethylammonium (TEA^+), and tetra-butylammonium (TBA^+) through CAHLM1 in oocytes indicated that the pore size was ~10 Å in diameter [9].

Ion channel	Maximum conductance (pS)	Permeability of organic molecules	Fluorescent dye uptake or PEGs	Pore size (Å)	Cutoff (Da)	Reference
CALHM1	24	$Ca^{2+} > > Na^{+} \sim K^{+} > Cl^{-}$ Weak permeability to divalent cations other than Ca^{2+} TMA^{+}, TEA^{+}, TBA^{+}, and ATP^{-4}	**Anionic:** LY^{-2}, $Alexa350^{-1}$, $Alexa488^{-2}$, and slight $Alexa594^{-2}$ **Cationic:** YO-PRO-1^{+2}	14	~ 800	[9, 23–26]
Cx	Cx45–57 Cx43–250 Cx40–129 Cx37–300 Cx32–90 Cx26 – 130	$Cs^{+} \geq Rb^{+} > K^{+} > Na^{+} \geq Li^{+}$ Ca^{2+} $Cl^{-} > NO^{3-}$ $TMA^{+} > TEA^{+}$ IP_3, $cAMP^{-}$, ADP^{-2}, and ATP^{-4}	EG, TriEG, DiEG, PEG 200, 300, and 400 **Anionic:** LY^{-2}, and CF^{-}, $Alexa350^{-}$, $Alexa488^{-2}$, $Alexa594^{-2}$, $Alexa350^{-1}$, $Alexa488^{-2}$, and $Alexa594^{-2}$ **Cationic:** YO-PRO-1^{+2}, EB^{+}, $DAPI^{+}$ and PI^{+2}	11–16	~ 800	[11–14, 27–38]
P2X2R	~30 at −100 mV or 10 at −50 mV	$K^{+} > Rb^{+} > Cs^{+} > Na^{+} > Li^{+}$ $PCa^{2+}/P_{Na}{}^{+} \sim 2.5$ $NMDG^{+}$ DMA^{+}, 2-MAE^{+}, and $Tris^{+}$	**Cationic:** YO-PRO-1^{+2} and FM1–43^{+2}	14–15.3	~ 700	[7, 15, 39–45]
P2X4R	9	$Na^{+} = K^{+}$ and Ca^{2+} ($PCa^{2+}/P_{Na}{}^{+} \sim 4.2$) $NMDG^{+}$	**Cationic:** YO-PRO-1 and EB^{+}	~16	~ 700	[39, 46–48]
P2X7R	Low permeability state with 9 and 13 High permeability with 423–445	Na^{+}, K^{+} and Ca^{2+} ($P_{Ca}{}^{2+}/P_{monovalentT} \sim 4.5$) $Tris^{+}$, $NMDG^{+}$, and $glutamate^{-}$	**Anionic:** LY^{-2}, CF^{-}, fura-2^{-5}, fluorescein $isothiocyanate^{-1}$, $fluorescein^{-1}$, and $calcein^{-4}$ **Cationic:** YO-PRO-1^{+2}, PI^{+2}, EB^{+1}, TO-TO-1^{+4} and POPO-3^{+4}	7–40	900	[17, 19, 22, 39, 49–56]
Panx-1	475 in 150 mM potassium gluconate and 550 in 150 mM KCl	$NO^{-3} > I^{-} > Br^{-} > Cl^{-} > F^{-} \gg aspartate^{-} \approx glutamate^{-} \approx gluconate^{-}$ Ca^{2+} ATP^{-4}	**Anionic:** $Alexa350^{-1}$ **Cationic:** YO-PRO-1^{+2} and EB^{+}	17	900	[21, 57–61]
pl-VDAC1	~400	Cl^{-} and ATP^{-4}	?	25–30	1000	[62–66]

Ion channel	Maximum conductance (pS)	Permeability of organic molecules	Fluorescent dye uptake or PEGs	Pore size (Å)	Cutoff (Da)	Reference
MAC	640 640	Cl^{-}, formate$^-$, acetate$^-$, propanoate$^-$, glutamate$^-$, pyruvate$^-$, methanesulfonate$^-$, gluconate$^-$, glucuronate$^-$, glucoheptonate$^-$, and lactobionate$^-$ ATP^{-4} and $MgATP^{2+}$	EG, glycerol, PEG 200, 300, 400, 600, 1000, 1540, and 2000 **Anionic**: Calcein^{-4} and fluorescein^{-1} **Cationic:** PI^{+2} partially	~30	1000	[10, 56, 67, 68]
TRPV1	~50 at −60 mV and ~ 100 at +60 mV	$Ca^{2+} > Mg^{2+} > Na^{+} > K^{+} > Cs^{+}$ 2-MAE$^+$, Tris$^+$, and NMDG$^+$	**Cationic:** YO-PRO-1^{+2}, FM1–43^{+2}, PI^{+2}, and EB^{+1}	19	900	[15, 56, 69–73]
TRPV2	101 pS in 140 mM NaCl symmetric at +90 mV, but lower at – 90 mV with a value of 28 pS	$Ca^{2+} > Mg^{2+} > Na^{+} \sim Cs^{+} \sim K^{+}$; $P_{Ca}{}^{2+}/P_{Na}{}^{+} = 2.94$; $P_{Mg}{}^{2+}/P_{Na}{}^{+} = 2.40$ NMDG$^+$	**Cationic:** YO-PRO-1^{+2}	12.3 in mutant TRPV2	~700	[74–77]
TRPV3	337	Ca^{2+}, Na^{+}, K^{+}, and NH^{4+} $P_{Ca}{}^{2+}/P_{Na}{}^{+} = 10$ NMDG$^+$	N/A	13.6	N/A	[78–81]
TRPV4	310 in hypotonic conditions	$Ca^{2+} > Mg^{2+} > K^{+} > Cs^{+} > Rb^{+} > Na^{+} > Li^{+}$ $P_{Ca}{}^{2+}/P_{Na}{}^{+} = 6–10$	N/A	13.8	~500	[82–85]
TRPA1	~100	$Ca^{2+} > Ba^{2+} > Mg^{2+} > NH^{4+} > Li^{+} > Na^{+} > K^{+} > Rb^{+} > Cs^{+}$ NMDG$^+$, MA$^+$, DMA$^+$, TriMA$^+$, and TMA	**Cationic:** YO-PRO-1^{+2}, FM1–43^{+2}, EB$^+$, and PI^{+2}	16	900	[8, 56, 86–90]
TRPM8	95	$P_{Ca}{}^{2+}/P_{Na}{}^{+}$, $P_{K}{}^{+}/P_{Na}{}^{+}$, and $P_{Cs}{}^{+}/P_{K}{}^{+}$ of 3.3, 1.2, and 1, respectively	**Cationic:** YO-PRO-1^{+2} and PO-PRO-3^{+2}	> 11	~ 700	[18, 87, 88, 91–94]
MTX- induced pore	466.9	High Ca^{2+} permeability $NH4^{+} > K^{+} > Na^{+} > Li^{+}$ NMDG$^+$	**Anionic:** fura-2^{-5} **Cationic:** EB$^+$, YO-PRO-1^{+2}, and POPO-3^{+4}	> 14	900	[17, 21, 95–100]
Ionomycin-induced pore	411–421	Ca^{2+}	**Anionic:** LW^{-2} **Cationic:** EB$^+$, PI^{+2}, and POPO-3^{+4}	> 14	900	[49, 101]

Table 1.
Biophysical properties and selectivity of pore-forming ion channels.

CALHM1 is involved in the mechanism of food flavors, such as sweet, bitter, and umami flavors. Stimulation in taste buds mediates the activation of G protein-coupled receptors, which recruit Ca^{2+} from the endoplasmic reticulum. An increase in intracellular Ca^{2+} induces depolarization of the plasma membrane through the TRP receptor, releasing adenosine triphosphate (ATP^{-4}) from the cell and mediating the purinergic activation responsible for sending information to the brain. CALHM1 knockout mice lost their stimulated sensitivity to sweet, bitter, and umami tastes and exhibited electrophysiological alterations. Surprisingly, these conditions decreased ATP^{-4} release, suggesting that ATP^{-4} passes through CALHM1. Some authors classify CALHM1 as a voltage-gated ATP^{-4} channel [9, 23–26]. Thus, this would indicate a pore diameter > 13 Å (**Table 1**).

CALHM1 induces the passage of highly anionic fluorescent dyes, Lucifer yellow (LW^{-2}), $Alexa350^{-}$, and $Alexa488^{-2}$, into the cytoplasm of oocytes; however, the dyes $Alexa594^{-2}$ and $Alexa633^{-2}$ have slight and no uptake, respectively. The last two fluorophores can be explained by their large diameter and molecular weight, which would already exceed the channel limit (Alexa 594^{-2} 760 Da/13.8 Å and Alexa 633^{-2}/>1000 Da) (**Table A2**). Thus, the pore diameter was associated with ~14 Å. One point in question is that the unit conductance of the channel is approximately 24 pS, which is relatively low compared with the pore diameter. According to the authors, molecular interactions at different sites could explain this phenomenon during the passage of ions through the channel lumen. However, new investigations are needed to understand this mechanism [9]. Highly divalent cationic dyes, such as YO-PRO-1^{+2}, have also been evaluated in assays for CALHM1 [102].

2.2 Connexins (Cxs)

Cxs are hexameric ion channels known to form gap junctions, mediating communication between the cytoplasm and allowing the passage of organic molecules such as ATP^{-4}, glutamate, and Ca^{2+} ions between the cytosolic compartments of adjacent cells. There are approximately 21 Cxs expressed in the human genome. The Cx is one of the oldest pore-forming channels studied and has similar biophysical properties to some large conductance channels described in this work [103].

Hemichannels are expressed in monomeric form, but they also form heteromeric channels, which may imply the biophysical characteristics of the channel. In this work, we describe the characteristics of homomeric Cxs [104]. The Cx family is large; however, pore-forming channels are generally associated with Cx45, Cx43, Cx40, Cx37, Cx32, and Cx26. Cx37 presents the highest conductance at ~300 pS, whereas Cx45, Cx43, Cx40, Cx32, and Cx26 exhibit conductance values of 57, 250, 129, 90, and 130 pS, respectively, in whole-cell patches [12, 27–30]. Although the conductance varies between channels, the selectivity for monovalent cations decreases in the order of $K^+ > Na^+ > Li^+ > TMA^+ > TEA^+$ and that for divalent cations such as Ca^{2+} [31, 32]. Inositol 1,4,5-triphosphate (IP_3) and cyclic adenosine monophosphate ($cAMP^-$) cross through Cx43, Cx40, Cx32, and Cx26 [33]. Whole-cell recordings of Cx43-transfected neuroblastoma 2A cells revealed the following sequence of cation conductance: $Cs^+ \geq Rb^+ > K^+ > Na^+ \geq Li^+ > TMA^+ > TEA^+$ [34]. Anions, such as Cl^-, $acetate^-$ and $glutamate^-$, and ATP^{-4} also pass through Cx43 [35]. Similarly, between cations and anions, Cx40 follows the same order of permeability for cations and $Cl^- > NO^{3-}$ and low glutamate permeability [36]. The permeability of ATP^{-4} through Cx43, Cx32, and Cx26 suggests that these channels may be related to the purinergic system. Cx43 stands out for its permeability to AMP, adenosine diphosphate (ADP^{-2}) [11].

Initially, the permeability of Cx37, Cx32, and Cx26 to neutral polyethylene glycol probes (PEGs) with diameters ranging from 4.4 Å for ethylene glycol (EG) to 17.8 Å (PEG 1000) was evaluated (**Table A1**). Surprisingly, Cx32 has a pore for capturing PEGs (EG, DiEG, TriEG, PEG 200, 300, and 400) and an estimated diameter of up to 11.2 Å, whereas Cx37 and Cx26 (EG, TriEG, DiEG, PEG 200) have size estimates of ~8 Å [12]. Interestingly, three anionic fluorescent dyes, Alexa350, $Alexa488^{-2}$, and $Alexa594^{-2}$, ranging in molecular weight from 350 to 760 Da and diameter from 5.4 to 13.8 Å, were studied in homotypic gap junctions. Cx45, Cx43, Cx40, Cx37, Cx32, and Cx26 are expressed in oocytes, in which only Cx43 and Cx32 mediate the uptake of all probes, with a slight increase in Cx32. On the other hand, the permeability of Cx26, Cx40, and Cx45 decreased from 20-fold for Cx26 to 45-fold for Cx40 and Cx45 between the smaller and larger probes. This suggests an estimated pore diameter of ~14 Å, whereas Cx37 permits the flux of two smaller probes, with an estimated pore size of ~11 Å. On this basis, a pore comparison was proposed as follows: Cx32 $\geq$ Cx 43 $\gg$ Cx26 $\geq$ Cx40 $\sim$ Cx45 $\geq$ Cx37. Owing to the greater permeabilization capacity observed for Cx43 and Cx32, the authors suggest that the pore diameter for both channels reaches 16 Å (**Table A2**) [11, 37].

Interestingly, the negative probes LY^{-2}, carboxyfluorescein (CF^{-}) and $Alexa350^{-}$, exhibited permeability for four Cxs in the following order: Cx43 > Cx45 > Cx26 > Cx40, whereas the positive probes ethidium bromide (EB^{+}) and Cx26 $\geq$ Cx43 $\approx$ Cx40 > Cx45 [14]. Despite the nonselectivity of the channels, this indicates that probable interactions between dyes and the pore structure should influence their permeability rate. Another interesting point is the functionality of the Cx channel according to the cell type in question or the species of origin of the channel. For example, the transfection of chick Cx45 into N2A neuroblastoma cells did not affect the permeabilization of CF- but rather that of dichlorofluorescein [38]. Similar findings were obtained in oocytes expressing human Cx26, which permit the flux of $DAPI^{+}$ but not to EB^{+}, whereas HeLa cells expressing human Cx26 permit the flux of EB^{+} [13].

Cx43 is one of the main proteins involved in high-conductance pore formation. Whole-cell current recordings in a planar lipid bilayer revealed an ionic conductance of 224 pS and a subconductance of up to 124 pS in voltage ranges between −100 and + 100 mV. Interestingly, the fusion of green fluorescent protein with Cx43 increased the electrophysiological properties of the pore. These properties are marked by conductance states ranging from 15 to 250 pS, with a higher probability of opening at 0 mV [105]. Cx43 leads the passage of cationic EB^{+}, propidium iodide (PI^{+2}), $DAPI^{+}$, and $YO\text{-}PRO{-}1^{+2}$ and anionic $Alexa350^{-}$, $Alexa488^{-}$, $Alexa594^{-}$, LY^{-2}, and CF^{-} fluorescent probes [14, 37, 106].

Interestingly, in the early 1990s, Cx43 pores were shown to depend on ATP^{-4} in mouse macrophages and strain J774 [28]. This was intriguing for the study of pores because Steinberg and collaborators established that ATP^{-4} is a pore mediator that allows the passage of dyes such as CF, LW^{-2}, and $Fura\text{-}2^{-5}$ [20, 107]. Soon after, other authors demonstrated that this effect was linked to purinergic P2 receptors, more specifically to P2Z (formerly the P2X7R nomenclature), which showed the same response patterns as those in Steinberg's study [107]. Therefore, Cx43 could function as a pore associated with P2X7R. However, our group ruled out this possibility on several occasions, in which both connexin hemichannel antagonists and knockout animals were unable to prevent ATP^{-4}-induced pore formation [49, 108, 109].

2.3 Purinergic receptor P2X2Rs

P2X receptors are channels activated by extracellular ATP^{-4} ($eATP^{-4}$) and belong to the purinergic system. Among the P2X receptors, pore formation involves P2X2R, P2X4R, and P2X7R, the latter being the most notable. P2X2R, P2X4R, and P2X7R desensitize more slowly to $eATP^{-4}$, whereas other receptors in the P2X family are quickly desensitized, a hypothesis that would explain why these highlighted receptors increase permeability to large molecules [39].

$K^+ > Rb^+ > Cs^+ > Na^+ > Li^+$ are the selectivity sequences for the influx of currents from P2X2Rs with monovalent cations. Outside-out patch recordings of HEK293 cells expressing P2X2R revealed a conductance of ~30 pS at −100 mV and internal rectification in symmetric solutions with 145 mM NaCl, but this conductance may decrease at more positive voltages of ~10 pS at −50 mV [40]. P2X2R is permeable to Ca^{2+} ($P_{Ca}{}^{2+}/P_{Na}{}^{+}$ ~ 2.5); in contrast, divalent cations tend to block the currents produced by the channel in the following order: $Mn^{2+} > Mg^{2+} > Ca^{2+} > Ba^{2+}$ [110].

Whole-cell recordings of HEK293 cells expressing P2X2R stimulated with $eATP^{-4}$ in N-methyl-D-glucamine ($NMDG^+o/Na^+i$) solution revealed that the external current decreased quickly, becoming internal at 5 s. Changes in the reversal potential increase permeability to $NMDG^+$ in the intracellular environment. The rate of pore expansion is proportional to the duration of agonist stimulation. This effect is evident through fluorescence microscopy, in which YO-PRO−1^{+2} showed greater intensity after 60 s of $eATP^{-4}$ application via P2X2Rs [41]. In addition to $NMDG^+$, P2X2Rs also pass other cations, such as dimethylammonium (DMA^+), 2-(methyl-amino)-ethanol (2-MAE^+) and tris(hidroximetil) aminometano ($Tris^+$) [7]. HEK293 cells transfected with clone P2X2R in pcDNA3.1 also exhibited increased FM1-43^{+2} uptake after stimulation with $eATP^{-4}$ for 2 min, and this effect was reversed by P2X2R inhibition [15]. The pore diameter of P2X2R was calculated to be approximately 14 Å because of organic cation permeability [7]. Using cryoelectron microscopy, Mio and collaborators were able to estimate a diameter of 15.3 Å [42].

The pore associated with P2X2R was shown to be intrinsic to the channel, allowing conformational changes in the remote cytosolic domain with subsequent dilation. Additionally, patch clamp and fluorescence assays did not reveal that Panx−1 participates in P2X2R pore opening, which suggests that pore formation occurs without secondary messengers or another protein-mediating pore opening [43]. In contrast, tubulin was shown to bind to the C-terminal loop of P2X2R, suggesting that the receptor interacts with proteins in the cell cytoplasm [44]. Colchicine is a potent alkaloid that inhibits cell cytoskeleton functions and dose-dependently inhibits YO-PRO−1^{+2} and EB^+ in oocytes or HEK293 cells transfected with rat P2X2Rs. However, the electrophysiological recordings did not affect the currents evoked by eATP-4 [45]. These findings support the hypothesis that P2X2R pore properties are affected by disruption of the microtubule cytoskeleton, suggesting a possible influence of a second messenger.

Interestingly, the whole-cell configuration in bi-ionic solutions with $NMDG^+$ present outside the cell and Na^+ inside ($NMDG^+o/Na^+i$) after initial exposure to $eATP^{-4}$ caused P2X2Rs to be more permeable to Na^+ than to $NMDG^+$ without affecting the membrane potential. Prolonged agonist treatment increased intracellular $NMDG^+$ and decreased the equilibrium potential. This increase in $NMDG^+$ is accompanied by the depletion of internal Na^+, with significant changes in the internal concentration, where Na^+ decreases from 140 mM to approximately 20 mM, and $NMDG^+$ accumulates at concentrations greater than 200 mM. From this point of view, it was hypothesized that P2X2Rs would be permeable to $NMDG^+$ without pore dilation [111, 112].

2.4 Purinergic receptor P2X4Rs

Compared with P2X2Rs and P2X7Rs, P2X4Rs are more rapidly agonist-activated receptors for $eATP^{-4}$. The outside-out mode of human P2X4R expressed in HEK293 cells results in a conductance of 9 pS at a current–voltage relationship > −100 mV [46, 113]. The permeabilities of the monovalent cations Na^+ and K^+ are relatively equal. Likewise, divalent cations, such as Cu^{2+} and Mg^{2+}, also inhibit the P2X4R-mediated current. Furthermore, P2X4 currents are potentiated by Zn^+ and ivermectin in combination with $eATP^{-4}$, increasing the channel response. Ca^{2+} permeability is greater than P2X2R ($P_{Ca}{}^{2+}/P_{Na}{}^{+} \sim 4.2$) permeability [39].

Like P2X2Rs, P2X4Rs, when exposed to $eATP^{-4}$ for a prolonged period, progress to a state of expansion, allowing the permeability of charged and high-molecular-weight molecules. Khakh and collaborators showed that, in the whole-cell configuration using oocytes expressing P2X4R channels, an initial current was induced due to stimulation by $eATP^{-4}$ at submillimolar concentrations. The initial current was followed by a slower current that began to develop within seconds and peaked at 150–300%. This second current after $eATP^{-4}$ application in $NMDG^+$ solution increased the $_{P}NMDG^+/_{P}Na^+$ ratio from 0.097 to 0.20, an effect that reversed after the interruption of the stimulus [47]. With respect to $NMDG^+$ permeabilization, the calculated pore diameter reached 9 Å. However, when the P2X4R pore was stimulated with $eATP^{-4}$ plus modular allosteric Zn(II) and ivermectin, the pore diameter increased by up to ~16 Å [48].

Interestingly, although the reversal potentials of P2X2R and P2X4R after $eATP^{-4}$ stimulation are very close, the permeability ratio of $_{P}NMDG^+/_{P}Na^+$ was significantly lower for P2X4R (0.16) than for P2X2R (0.29) [41]. This finding is comparable to the low fluorescence intensities of $YO\text{-}PRO\text{-}1^{+2}$ and EB^+ uptake observed in the studies of Khakh and Marques-da Silva, respectively. The signal intensity was greater for P2X2Rs than for P2X4Rs. Both studies showed different experimental and methodological conditions (transfection in *Xenopus laevis* oocytes with wild-type P2X2R/P2X4R and HEK293 cells transfected with rat P2X2R and human P2X4R), but the prevalence of greater permeability by P2X2R persists in both situations [45, 47]. This possible phenomenon between receptors may be associated with the dye's interaction with the channel's structure or with the ionic conductance, which is more significant in P2X2R than in P2X4R (**Table 1**).

2.5 Purinergic receptor P2X7Rs

P2X7R is the most mysterious and promising purinergic receptor. Interestingly, higher concentrations of $eATP^{-4}$ are required for activating these channels (EC_{50} > 100 μM), whereas P2X2 and P2X4 require lower concentrations. The permeability of P2X7R to monovalent cations is comparable to that of other P2Xs, and P2X7R has considerable permeability to Ca^{2+} ($P_{Ca}{}^{2+}/P_{monovalent} \sim 4.5$). However, P2X7R currents are significantly inhibited by divalent cations in the rank order of $Cu^{2+} > Cd^{2+} = Zn^{2+} > Ni^{2+} >> Mg^{2+} = Co^{2+} > Mn^{2+} > Ca^{2+} = Ba^{2+} >> Sr^{2+}$ [39].

P2X7R is a good example of a target involved in inflammation, and the pore phenomenon is associated with cell death and activation of nod-like receptor 3 (NLRP3), which results in the activation of a protein cascade up to caspase−1 and subsequent maturation of cytokines such as IL − 1β and IL − 18 [114, 115]. However, pore opening is also associated with an increased effect of charged drugs that do not cross the cell membrane. Our group demonstrated through photodynamic therapy in strain J774 that the cytotoxic effect of the methylene blue+ compound was enhanced in combination with $eATP^{-4}$, suggesting the opening of pores and consequent internalization of the drug [116]. Similarly, other authors have investigated this topic in the context of cancer, in which the drug doxorubicin synergistically increased the sensitivity of Yac – 1 lymphoma cells expressing P2X7R through $eATP^{-4}$ stimulation [117].

Over many years of study, two theories have been discussed in the literature to explain the phenomenon of high-conductance pores in P2X7Rs. The first is that after activation of the receptor, the ion channel gradually dilates according to the time of exposure to the agonist (increase in intrinsic permeability). In contrast, the second is based on the presence of second messengers and associated proteins. The P2X7R dilation theory has been reported in several papers. This conformational change leads to the dilation of the integral pore of P2X7R, which initially forms a channel permeable to small cations and subsequently permeable to large cations or anions, including fluorescent dyes, characterizing a nonselective pore with high conductance [50]. $eATP^{-4}$ mediates two types of single-channel currents mediated by P2X7R, characterized by short- and long-lasting open states and slope conductance approaching 9 and 13 pS, respectively, at −60 mV. P2X7R activation led to $Tris^{+}$ and $NMDG^{+}$ uptake. The constant expansiveness of the channel diameter was suggested to increase from 7 to 40 Å [51], and gradual dilation generates YO-PRO-1^{+2} uptake [52].

CF^{-}, LY^{-2-}, and Fura-2^{-5} also pass through P2X7R but not through Trypan $blue^{-4}$ or Evans $blue^{-4}$. The last two dyes exceed 900 Da. Therefore, the pore permeability limit is estimated at 900 Da [107]. The potent agonist $BzATP^{-4}$ induced THP – 1 cells expressing P2X7R to take up EB^{+}, YO-PRO–1^{2+} and the tetravelent POPO-3^{+4} cationic dye [17]. Interestingly, Cankurtaran-Saya showed TOTO-1^{+4} uptake, which is very close to the limit of 900 Da. Interestingly, in this work, the transfection of rat P2X7Rs into HEK293 cells led to YO-PRO-1^{+2} and TOTO-1^{+4} uptake but not anionic LY^{-2} or $calcein^{-4}$ uptake, independent of the increase in intracellular Ca^{2+} concentration, whereas only LY^{-2} was absorbed in a Ca^{2+}-dependent manner. In RAW264.7 macrophages, all dyes, regardless of their charge, are absorbed independently of Ca^{2+}. These authors concluded that the different permeabilities of P2X7Rs to both cations and anions may depend on the activation stimulus [19].

It has been demonstrated that variations in the constitution of amino acids present in the channel lumen can lead to depletion of the functionality of membrane permeability to the cationic dye EB^{+} but not to the anionic dye fluorescein $isothiocyanate^{-}$. These mutations also increase the permeability of P2X7R to Cl^{-} rather than cations, which is the opposite of its basal function [22]. Regarding the structure of the channel, the C-terminal tail plays a fundamental role in the dilation of the receptor. Unlike other P2Xs, P2X7R has an elongated C-terminal tail. Changes in the second transmembrane domain of the receptor (partially linked to the C-terminal tail) depleted the uptake of the cationic dyes EB^{+}, $PI^{+2,}$ and YO-PRO-1^{+2} [53]. Surprisingly, Jiang and coauthors deleted an 18-amino acid cysteine-rich segment in the intracellular region of P2X7R and observed impaired $NMDG^{+}$ uptake and no gradual process of channel dilation. However, YO-PRO-1^{+2} uptake remained intact. These data conflict with previously mentioned data, suggesting two different mechanisms between $NMDG^{+}$ and YO-PRO-1 uptake though P2X7R [54].

In contrast, an interesting study from 2017 indicated that pore opening depends on the lipid constitution of the plasma membrane and is independent of the C-terminal domain. These authors reported that P2X7R activity is facilitated by phosphatidylglycerol and sphingomyelin and impaired predominantly by cholesterol through direct interactions with the transmembrane domain. Additionally, YO-PRO-1^{+2} had better uptake in HEK293 cells expressing P2X7R when the cholesterol-scavenging agent methyl-β-cyclodextrin (MβCD) was used [118].

Oocytes transfected with rat P2X7R maintained ion channel activity, reversing the current evoked by $eATP^{-4}$ or $BzATP^{-4}$, but there was no uptake of intracellular YO-PRO-1^{+2} [119]. Similarly, P2X7R human lymphocytes expressed in oocytes are unable to exhibit a pore-associated cellular permeabilization phenotype [120]. These data first suggest that the degree of permeabilization is correlated with the cell type in question and may cause changes in the functional phenotype of P2X7Rs. Second, the activation of P2X7R leads to a signaling cascade that can promote the activation and recruitment of second messengers, inducing the formation of pores through another component, according to previous publications [49].

Given these conflicting findings, the theory of an intracellular mechanism, independent of dilation, involves the study of P2X7R pores in other horizons. Previous studies have shown that the conductance of peritoneal macrophages and J774 strain macrophages increased by up to 409 pS after stimulation with $eATP^{-4}$ in cell-attached patches, increasing $Tris^{+}$, $NMDG^{+}$, and $glutamate^{-}$ uptake, in which temperatures above 30°C favor pore opening. These authors proposed that there is a need for second messengers that were not initially identified [55, 121]. Our group, which used whole-cell and cell-attached configurations combined with fluorescence, revealed the possible influence of second messengers and the involved intracellular signaling. The biophysical properties of P2X7R-associated pores in peritoneal macrophages and 2BH4 thymic epithelial cells are similar, with pore conductance ranging from 423 pS to 445 pS in 2BH4 cells, similar to previous findings [55]. Furthermore, higher Ca^{2+} concentrations in the micropipette solution induced strong EB^{+} uptake across the membrane. When combined with the Ca^{2+} ionophore ionomycin at 1 μM, $eATP^{-4}$ at a concentration of 25 μM, which is considered ineffective for pore induction, was able to take up EB+ in the cells, a condition that was affected when the Ca^{2+} was removed. Thus, Ca^{2+} is a second messenger of pore formation [49].

Another mechanism involved is the participation of intracellular proteins in generating P2X7R pores. SB202190, a selective p38 mitogen-activated protein kinase (MAPK) inhibitor, strongly inhibited $BzATP^{-4}$-stimulated YO-PRO-1 + 2 uptake, with an IC_{50} value of 75 nM, whereas MEK kinase and ERK2 did not cause pore blockage in THP-1 cells. Inhibitors of caspase-1 YVAD, caspase-3 (DEVD) and PKC (Ro-31-7549 and Ro-31-8220) also block $BzATP^{-4}$-evoked YO-PRO-1^{+2} uptake, with IC_{50} values of 8, 3.2, 5.2 and 3.5 μM, respectively, which are less potent than those of p38 [122]. Our group also demonstrated the importance of MAPKs in opening pores. Similarly, SB – 203,580, a p38 MAPK blocker, and PD-98059, an MKK inhibitor, potently inhibited ATP-induced dye uptake and single currents in 2BH4 cells and peritoneal macrophages, with IC_{50} values of 75, 104, 88, and 123 nM, respectively. Therefore, MAPKs and caspases play intrinsic roles in pore opening [49].

Interestingly, electrophysiological studies by Marques da Silva et al. revealed that in oocytes transfected with rat P2X2R and P2X7R receptors, responses related to $eATP^{-4}$-evoked currents were not affected by the microtubule disruptor colchicine. However, YO-PRO-1^{+2} uptake was potently inhibited after $eATP^{-4}$ stimulation, suggesting the inhibition of pore formation by these receptors. Furthermore, dye uptake by $eATP^{-4}$ stimulation alone at P2X7R was significantly more responsive and reproducible than that at P2X2R, likely due to the P2X7R pore-associated mechanisms discussed previously [45]. On the basis of the physicochemical and conductance properties of the pore-forming purinergic channels, the fluorescence intensity corresponded to the following order: P2X7R > P2X2R > P2X4R.

The real mechanism of pore opening by second messengers is uncertain, although data indicate the presence of mainly Ca^{2+} in this process. One discussion in this context is whether P2X7R has some protein unit that, after its activation, generates a pore in the membrane and increases its permeability. This phenomenon was attributed to the high-conductance proteins known as Cx and Panx–1. Furthermore, these proteins can conduct $eATP^{-4}$, which is linked to purinergic signaling and the high conductance pore associated with P2X7R [28]. Our group has ruled out the possibility that Cxs are associated with P2X7R pores [49, 108, 109].

Among these pores, Panx–1 was found to be related to P2X7R, mainly due to the very similar characteristics between the pores, such as conductance and nonselectivity. This finding is consistent with a study published by Pelegrín, in which they used HEK293 cells expressing rat P2X7R, the macrophage lineage J774, and pulmonary alveolar macrophages and reported that general hemichannel antagonists and Panx-1 RNA interference inhibited the uptake of $YOPRO\text{-}1^{+2}$ or EB^{+} after P2X7R activation by 60–100% [123, 124]. The P2X7R/Panx-1 interaction axis could suppress membrane currents from oocytes expressing rat P2X7R or coexpressing the receptor with human Panx-1 after $eATP^{-4}$ stimulation. However, some publications conflict with these results. Data from the same HEK293 cell line revealed that the biphasic current induced by P2X7R is independent of Panx-1 activation. In agreement with these data, our group reported that RNA interference in murine peritoneal macrophages did not inhibit the currents or dye uptake of $eATP^{-4}$ induced in electrophysiological and fluorometric experiments [49, 109, 125]. The Panx-1 topic is used for more information.

In addition to Panx-1 and Cx, our group associated other pore-forming channels that, because of their similar characteristics to P2X7R, such as conductance and pore cutoff, could be linked to a large complex of channels [21, 126]. In this sense, the effects of TRPV1, TRPA1, MAC, and pl-VDAC1 on the uptake of cationic and anionic dyes in astrocytes induced by $eATP^{-4}$/P2X7R and its antagonists were evaluated. Surprisingly, TRPV1, TRPA1, and MAC were the only large conductance channels whose fluorescence decreased after $eATP^{-4}$ stimulation. TRPV1 and TRPA1 inhibited only the cationic dyes PI^{+2} and EB^{+} (75 and 60%, respectively), whereas MAC inhibited the anionic dyes fluorescein and $calcein^{-4}$ (~65%) and partially inhibited PI^{+2} (20%). These results suggest that these channels stimulate purinergic signaling. However, when stimulated by their own agonists in whole-cell or inside-out patch clamp recordings in cells lacking P2X7Rs, both receptors evoke currents, suggesting P2X7R-independent functionality [56]. More studies involving this signaling pathway are needed.

2.6 Pannexin 1 (Panx-1)

Like Cx and CALHM1, Panx-1 channels are hexameric proteins. In addition, Panx-1 present in the plasma membrane consists of ion channels but not functional gap junction channels, which differ from connexins [127, 128]. The Panx-1 channel is often associated with the P2X7R pore-forming unit. However, some evidence indicates controversies related to this statement, as discussed in the topic of P2X7R [109].

Whole-cell experiments revealed that externally rectifying Panx-1 currents are activated at voltages greater than ~10 mV. The authors suggested that Panx-1 forms an anion-selective channel in HEK293 cells in the following order: $NO^{-3} > I^{-} > Br^{-} > Cl^{-} > F^{-} \gg aspartate^{-} \approx glutamate^{-} \approx gluconate^{-}$ [57]. However, Panx-1 leads to the uptake of cations, such as Ca^{2+} [21]. Bao and colleagues initially described Panx-1 as a protein capable of conducting ATP^{-4}. Inside-out analysis of human Panx-1 expressed in oocytes revealed a large unitary conductance of 475 pS in 150 mM potassium gluconate and 550 pS in 150 mM KCl. Under mechanical stress conditions, ATP^{-4} release via Panx-1 increases with membrane depolarization in a high-K^{+} solution [58]. Thus, ATP^{-4} conductivity through the Panx-1 channel by strong depolarization leads to its association with the purinergic system [59].

Panx-1 is permeable to high-molecular-weight fluorescent compounds [129]. The question of the high-conductance pore-forming unit of P2X7R has further added to the mystery behind purinergic receptor studies. Experiments on HEK293-P2X7R cells, J774 cells, and lung alveolar macrophages revealed that Panx-1 RNA interference inhibited $eATP^{-4}$-mediated YO-PRO-1^{+2} or EB^{+} uptake by 60–100%. Interestingly, inflammation-related signaling promoted by P2X7R activation was inhibited [123, 124]. More recent authors point to the joint activity of Panx-1 and P2X7R in cell death and the expression of proinflammatory mediators induced by *Clostridioides difficile* toxins in mediating caspase-3/7 activation [130]. Furthermore, a cryo-EM structural study of WT Panx-1 demonstrated that the channel has a wide pore diameter of 17 Å and an estimated pore size similar to that of the P2X7R pore [60].

However, Panx-1 anion selection is also controversial. Although the inhibition of Panx-1 reduces the uptake of cationic dyes, Sahu and collaborators reported that Panx-1 does not capture the dyes $DAPI^{+}$, EB^{+}, and PI^{+2} but does mediate $Alexa350^{-}$ uptake [61]. Our group, via whole-cell patch clamp, revealed that RNA interference targeting Panx-1 or Panx-1 inhibitors did not affect macroscopic P2X7R currents stimulated by $eATP^{-4}$ in murine peritoneal macrophages. This result was similar to data previously described by Yan and colleagues, in which the biphasic current response induced by $eATP^{-}4$ was found to be independent of Panx-1 activation [109, 125]. Furthermore, the absence of an influence of Panx-1 on $eATP^{-4}$/P2X7R-induced fluorescent dye uptake in peritoneal macrophages was also observed in our study [109].

Another interesting finding is that Ma et al., who studied HEK293 cells expressing Panx-1 but not P2X7R, quickly and reversibly reduced the Panx-1 current via the triphosphate nucleotides ATP^{-4}, GTP, and UTP through a mechanism independent of purine receptors. On the basis of these data, these authors proposed two populations of Panx-1: the first population associates within the P2X7R protein complex, whose conformation does not allow direct binding of ATP^{-4}, and the second population does not associate with P2X7R and is available for direct ATP^{4} inhibition [131]. Such a theory could explain the controversies between the permeabilization of fluorescent dyes and ionic currents. Finally, P2X7R pore theories may also influence this mechanism [43].

2.7 Plasma voltage-dependent anion channel 1 (pl-VDAC1)

Voltage-dependent anion channels (VDACs), also called porins, are pore-forming proteins first described in the outer mitochondrial membrane that play a relevant role in mitochondrial ATP^{-4} transport [132]. Apoptotic conditions can activate VDAC1, which is believed to be linked to mitochondrial cytochrome c release [133]. Some authors claim that pl-VDAC1 is a unit composing the MAC. However, this subject is controversial. See this work for more information [62].

High-conductance pores have been evaluated in different cell types, in which permeability via VDAC1 is preferentially increased by perturbations in the electrical potential of the cell. This channel is highly permeable to Cl^{-} and regulates cell volume [132]. Inside-out data from HT22 neuronal cells revealed a pore with 397 pS conductance. [133]. Similar conditions in rat astrocytic cell lines in external patch mode resulted in a maximum conductance of 401 pS, with intermediate subconductances ranging between 70, 125, 230, and 290 pS [63]. Initially, this large conductance pore was described as cylindrical, with a diameter of approximately 25--30 Å [64]. The crystallized structure of VDAC1 in lipid micelles showed pore diameter conformations measuring 27 Å, whereas the diameter of the narrowest point, which contains the α-helix, was measured at 14 Å [65].

Cell swelling leads to the activation of pl-VDAC1 currents and the release of ATP^{-4} into the extracellular environment in C127i cells. The concentration of intracellular ATP^{-4} is 2.5 mM, and electrophysiological measurements have shown that VDAC1 channels can transport ~10^5 ATP^{-4} s^{-1} at −30 mV [66]. Other cellular studies have indicated that fibroblasts from VDAC1 knockout mice release less ATP^{-4} than those from wild-type mice do in response to environmental changes. The same occurred in cells exposed to hypotonic luminal conditions [134]. Furthermore, the nucleotide-binding site at the C-terminus of mitochondrial VDAC1 presents a C-terminal sequence similar to that of pl-VDAC1 [135]. Therefore, pl-VDAC1 may be related to the purinergic system.

Even with these possibilities, our group performed a study in which astrocytes were stimulated to investigate the participation of pore-forming channels in the $eATP^{-4}$/P2X7R pathway. However, the use of a pl-VDAC1 antagonist did not prevent the uptake of fluorescent dyes by cells, ruling out its influence as a pore associated with P2X7Rs [56]. the use of fluorophores to measure the physicochemical properties of the pl-VDAC1 pore are not informative.

2.8 Maxi anion channels (MACs)

The MAC was first described as selectively permeable to the Cl^- anion. This channel opens immediately after the voltage step and closes 2.6 s after the step to −21 mV. Ionic conductance was measured at a value of approximately 430 pS, suggesting a high conductance channel property [136, 137]. In the context of the membrane potential, the maximum probability of opening occurs at approximately 0 mV but closes when the voltage exceeds the range of ±15 to ±30 mV.

Owing to the extracellular concentration of Cl^-, the conductance can increase to 600 pS, depending on the cell type studied. The single-channel conductance in L6 myoblasts was 640 pS 581 pS in T lymphocytes and 617 pS in frog skeletal muscular sarcomas. Furthermore, electrophysiological studies using different approaches have revealed subconductances of the MAC ranging from 5 to 50, 100, 150, and 200 pS [67, 126].

Interestingly, hypotonic, ischemic, hypoxic, and swelling conditions activate the MAC and the pore with a conductance close to 200 pS. Whole-cell patch clamp experiments revealed that replacing Cl^- with glutamate in the bath solution changed the reversal potential to −33.8 mV, with a permeability ratio of glutamate/Cl^- of 0.20 in neonatal rat cardiomyocytes. Furthermore, with the replacement of ATP^{-4} or $MgATP^{-2,}$ the reversal potential changed from 0 mV to −14.8 mV or – 30.9 mV, respectively. This effect induces the permeability of these compounds through the MAC with a $P_{ATP\text{-}4}/P_{Cl\text{-}}$ ratio of 0.12 and a P_{MgATP}/P_{Cl}^- ratio of 0.16. Similar effects were observed in C127 cells under ATP^{-4}-induced swelling conditions [10, 66, 68, 138, 139].

The sides indicated an asymmetry of the two-channel pore entrances and a medial constriction of a minimal pore size of 15 Å in diameter. These measurements confirmed the permeability of ATP^{-4}, $MgATP^{-2}$, formate, acetate, propanoate, glutamate, pyruvate, methanesulfonate, gluconate, glucuronate, glucoheptonate, and lactobionate (**Table A1**). These data suggest that MAC may be linked to purinergic and glutamatergic system signaling. Furthermore, a bilateral symmetrical application using polyethylene glycol molecules (EG, glycerol, PEG 200, 300, 400, 600, 1000, 1540, and 2000) of different sizes results in a pore size of ~30 Å in diameter [10, 68].

The influence of MAC on dye uptake in astrocytes induced via $eATP^{-4}$/P2X7R was evaluated, in which the dyes $calcein^{-4}$ and $fluorescein^{-4}$ were inhibited by the antagonists okadaic acid, $GdCl_3$, and arachidonic acid by 30% and 65%, respectively. Interestingly, MAC also modestly inhibited PI^{+2} by approximately 20%, suggesting cationic permeability. However, inside-out experiments in L6 cells without P2X7R, $eATP^{-4}$ and $BzATP^{-4}$ did not induce currents through the MAC, indicating P2X7R-independent functionality [56].

Another activation mechanism addressed would be opening by tyrosine phosphatases. Channel opening recorded by whole-cell patch clamp was abolished after treatment with protein tyrosine phosphatase inhibitors but not serine/threonine or protein kinases in C127 cells. Furthermore, knockout cells for protein tyrosine phosphatase did not allow the channel to open, and this effect was reversed in tests on wild-type cells [140]. Therefore, intracellular signaling appears to be involved in the opening and activation of the MAC.

2.9 Transient receptor potential vanilloid 1 (TRPV1)

TRPV1 is a thermoreceptor and is undoubtedly the most studied TRP channel in the literature. This receptor is directly involved in inflammation but is expressed mainly in neurons and is necessary for the pathophysiology of pain. The main agonist of TRPV1 is capsaicin (EC_{50} ~ 700 nM), the active component of peppers; on the other hand, this receptor can also be activated by hot temperature (45°C) protons, variations in the pH range of <6.5 and the capsaicin analog resiniferatoxin (EC_{50} ~ 40 nM), which is an ultrapotent agonist [141–144].

TRPV1 induces Ca^{2+} uptake rather than Na^{+} ion uptake. TRPV1 selectivity affinity was in the following order: $Ca^{2+} > Mg^{2+} > Na^{+} > K^{+} > Cs^{+}$. These dynamic changes in ionic permeability have been linked to mechanisms of pain hypersensitivity. The pore associated with TRPV1 has some similarities with that associated with P2X7R [69]. In DRG neurons and oocytes expressing TRPV1, the single-channel conductance at −60 and + 60 mV was ~50 and ~ 100 pS, respectively [70].

Initially, Chuang and colleagues proposed, through whole-cell measurements in HEK293 cells, prolonged stimulation with capsaicin-induced intracellular uptake of $NMDG^{+}$, and this effect was reversed upon withdrawal of the agonist. TRPV1 pore opening occurs exceptionally through channel dilation in a hemichannel-independent process. The permeability through the pores gradually increases for high cations, followed by 2-MAE^{+}, $Tris^{+}$, and $NMDG^{+}$. Interestingly, TRPV1 is described as N-glycosylated, and this factor is one of the main determinants of desensitization and ionic permeability evoked by capsaicin. Therefore, changes in N-glycosylation result in disturbances in ionic permeability and receptor functionality [145]. It was possible to evaluate the initial diameter of the pore at 10.1 Å, whereas prolonged stimulation with capsaicin increased the pore diameter to 12.3 Å for 3 minutes [71]. However, structural studies revealed a size of 19 Å for the TRPV1 channel, as determined by single-particle cryo-electron microscopy [72].

Interestingly, when YO-PRO-$1^{+2 \text{ was used}}$, intracellular uptake through TRPV1 was observed to occur in a time- and concentration-dependent manner [71]. Similarly, HEK293 cells transfected with the TRPV1 clone in pcDNA3.1 took up the FM1–43^{+2} dye with stimulation by capsaicin or 47°C thermal stimulus conditions after 3 minutes of induction [15]. Using MβCD in Chinese hamster ovary cells, the influence of cholesterol depletion on TRPV1 permeabilization was assessed with voltage ramps measured during 1 min of capsaicin application. A decrease in the cellular cholesterol content resulted in a significant decrease in pore dilation, which resulted in a decrease in $NMDG^+$ permeabilization of approximately 70%. After 20 s of agonist stimulation, YO-PRO-1^{+2} dye uptake was reduced by approximately 50% in cells treated with MβCD. However, such effects did not affect ion channel functionality, suggesting that membrane lipid constitution influences pore function [73].

The pores of TRPV1 and P2X7Rs have already been targeted due to their similarities, such as tissue expression and cellular communication [21, 126]. $eATP^{-4}$ has been reported to potentiate the activity evoked by capsaicin or proton currents in HEK293 cells expressing TRPV1 in a PKC-dependent manner [146]. Stimulation of dye uptake via $eATP^{-4}$/P2X7R showed that TRPV1 influences the uptake of PI^{+2} and EB^+, with both the TRPV1 antagonist ruthenium red and AMG-9810 reducing fluorescence in astrocytes by 40% and 75%, respectively. In contrast, the anionic compounds $calcein^{-4}$ and fluorescein were not inhibited by the TRPV1 antagonists, reinforcing their cationic selectivity. In the present study, whole-cell patches of cells that did not express P2X7Rs emitted TRPV1 currents after capsaicin stimulation but not after stimulation with $eATP^{-4}$ or $BzATP^{-4}$, suggesting that the functionality of TRPV1 is independent of P2X7Rs for ionic functions. Future studies that address this signaling pathway are needed [56].

The TRPV1 pore was proposed to facilitate the entry of drugs into cells and enhance the effect being studied in vivo. TRPV1 mediates the uptake of the gentamicin–Texas Red conjugate in the Madin–Darby canine kidney distal tubule cell line via an endosome-independent pathway and in hair cells of the inner ear [147, 148]. Rat dorsal root ganglion neurons showed that gabapentin uptake via capsaicin/TRPV1 was approximately 500x faster than that via amino acid transporters, significantly enhancing analgesic effects [149]. QX-314 (N-ethyl-lidocaine) is a small membrane-impermeable cation derived from lidocaine that acts as an inhibitor of voltage-activated Na^+ channels and produces long-lasting local anesthesia in animal models in vivo [150]. The expression of TRPV1 in HEK293 cells increased QX-314 permeabilization after capsaicin stimulation. However, capsaicin treatment in conjunction with QX-314 has previously shown neurotoxic effects in rats [151, 152]. Therefore, these data suggest that the toxicity of QX-314 may occur via a TRPV1-dependent mechanism. Lidocaine was shown to generate a TRPV1-dependent increase in Ca^{2+} and inward currents in DRG neurons [153]. In turn, 0.2% QX-314 administered with lidocaine had prolonged antinociceptive effects on pain-relieving mice and rats but not on TRPV1 knockout animals. These authors reported that the effect of lidocaine, rather than capsaicin, results in sustained regional analgesia without harmful effects [154].

2.10 Transient receptor potential vanilloid 2 (TRPV2)

TRPV2 is a thermoreceptor with approximately 50% homology to TRPV1. Temperatures above the harmful limit (>52°C) activate TRPV2. Like TRPV1, TRPV2 has greater selectivity for Ca^{2+} transport than other mono- and divalent cations do, following the order $Ca^{2+} > Mg^{2+} > Na^+ \sim Cs^+ \sim K^+$; $P_{Ca}{}^{2+}/P_{Na}{}^+ = 2.94$; $P_{Mg}{}^{2+}/P_{Na}{}^+ = 2.40$. Inside-out recordings of rat TRPV2 expressed in HEK293 cells revealed conductance in the main state of 101 pS in 140 mM NaCl, which was symmetric at +90 mV, but lower conductance at – 90 mV, with a value of 28 pS [74, 75].

Compan and colleagues reported that, after the TRPV2 gene was silenced in THP-1 cells, there was a significant reduction in the intracellular uptake of YO-PRO-1^{+2}.

Therefore, TRPV2 may have sufficient permeability to cross molecules up to 700 Da [76]. On the basis of the stipulated size of YO-PRO-1^{+2} (~10.6 Å), the pore size of TRPV2 may be similar to this value. The Zubecevic group elegantly demonstrated the crystallized structure of rabbit TRPV2, and these data were similar to those of TRPV1. Mutations in TRPV2 allowed the construction of the receptor by the binding of resiniferatoxin (a TRPV1 agonist), in which the estimated pore size increased by approximately 12.3 Å. Inside-out recordings of currents evoked by NMDG^{+} after resiniferatoxin stimulus via TRPV2 with a reversal potential range > +70 mV. Furthermore, these authors also showed that YO-PRO-1^{+2} uptake is mediated by TRPV2 over time, suggesting a process of pore dilation [77].

Interestingly, in glioma and multiple myeloma cancer cells, the cytotoxic effects of the anticancer drugs temozolomide, carmustine, doxorubicin, and bortezomib are amplified following TRPV2 activation by the agonist cannabidiol [155, 156]. Therefore, these effects may be related to the increased permeabilization of these receptors by these drugs, which act synergistically.

2.11 Transient receptor potential vanilloid 3 (TRPV3)

TPRV3 is a thermoreceptor with a temperature range of 31–39°C. TRPV3 has significant permeability to Ca^{2+} and other cations, such as Na^{+}, K^{+}, and NH^{4+} [78].

The first evidence of increased TRPV3 permeability was described by Chung and colleagues, in which TRPV3 biphasic currents were chemically induced or thermodependently stimulated in whole-cell clamps. These currents are similar to those previously described for P2Xs and TRPV1 receptors. Recombinant TRPV3 in keratinocytes facilitates NMDG^{+} uptake through prolonged exposure to stimuli, suggesting dilation. The permeability ratio in the second current was $P_{NMDG}{}^{+}/P_{Na}{}^{+}$ = 0.35, approximately five times greater than that in the first current (~0.07).

In contrast, Ca^{2+} permeabilization was greater at the first $P_{Ca}{}^{2+}/P_{Na}{}^{+}$ = 10, more than 4.5 times greater than at the second ~2.2 [79]. Inside-out recordings demonstrated that temperature variations directly influence the increase in conductance by up to 3-fold relative to that of TRPV1. For example, human TRPV3-transfected HEK293 cells were activated at 39°C by the agonist 2-APB, with a reversal potential of ~0 mV and inward and outward slope conductance of 337 and 256 pS, respectively. At room temperature (~25°C), the inward and outward slopes are 201 and 147 pS, respectively, with similar reversal potentials [80].

Recently, Lansky and coauthors reported that TRPV3 can act as both tetrameric and pentameric assemblies in lipid bilayers. In fact, the pentameric state is not a common phenomenon, but this stage is reversible and transient, with the tetrameric state being interconvertible with the pentameric state on the time scale of seconds to minutes. Surprisingly, using the cryoelectron microscopy technique, the authors observed different pore sizes depending on the structure in which the receptor is located. For example, the pentameric form has an estimated average diameter of ~13.6 Å, approximately 1.5 times larger than that of the tetrameric form (~8.9 Å) [81]. However, further studies using large molecules are needed.

2.12 Transient receptor potential vanilloid 4 (TRPV4)

TRPV4 acts as a thermoreceptor, but unlike the first TRPVs, its basal function occurs at moderate temperatures ranging between 24°C and 32°C, but it can also be activated at more elevated temperatures, such as 38°C. TRPV4 induces divalent cation uptake rather than monovalent cation uptake in the following order: Ca^{2+} > Mg^{2+} > K^{+} > Cs^{+} > Rb^{+} > Na^{+} > Li^{+}, P_{Ca}^{2+}/P_{Na}^{+} = 6–10 [82]. At 38°C, the agonist 4αPDD activates identical single-channel currents, and conductance across the channel varies at positive potentials (90–100 pS at +60 mV) and decreases at negative potentials (60 pS at −60 mV) in cell-attached recordings of HEK293 cells expressing wild-type TRPV4 [83]. Inside-out data in CHO cells under hypotonic conditions at +80 mV, the conductance is up to 3 times greater, with a range of 310 pS [84].

A recent study evaluated the structure of human TRPV4 in the presence of 4α-PDD, which induced significant changes in the channel gate estimated at up to 13.8 Å in diameter [85]. However, the only reports related to pore permeability involve the uptake of aminoglycosides by TRPV4. In murine distal renal tubule cells, TRPV4 activation increased gentamicin-Texas Red uptake in a manner dependent on extracellular Ca^{2+} removal, and this effect was abolished in cells expressing mutant TRPV4 [157]. Replacing extracellular Na^{+} with $NMDG^{+}$ abolished inward currents in HEK293 cells expressing wild-type TRPV4 [83]. Whole-cell recordings with voltages ranging between −80 and + 80 mV in smooth muscle cells from cerebral arteries confirm previous findings, suggesting that $NMDG^{+}$ does not pass into the intracellular compartment [158]. These findings include chemical induction (agonists) or heat stimuli incapable of inducing TRPV4 permeation to $NMDG^{+}$. Studies on the uptake of cationic and anionic fluorescent dyes must be performed during chemical or heat activation.

2.13 Transient receptor potential ankyrin 1 (TRPA1)

TRPA1 is a crucial thermoreceptor for low temperatures. Like other TRPs, this receptor is a nonselective mono- or divalent cation channel. As a cold sensor, TRPA1 is activated in neuronal cells and responds to cold temperatures with a limit of approximately 17°C [86].

TRPA1 leads to Na^{+}, K^{+}, and Ca^{2+} uptake, but when activated by the agonist allyl isothiocyanate (AITC), this agonist significantly increases intracellular Ca^{2+} levels. The permeability of smaller cations with different valences followed the order of Ca^{2+} > Ba^{2+} > Mg^{2+} > NH^{4+} > Li^{+} > Na^{+} > K^{+} > Rb^{+} > Cs^{+}. Outside-out patches revealed an increase in $NMDG^{+}$ permeability by turning the channel to a dilated state in rat TRPA1-expressing HeLa cells after agonist stimulation of ~4.4-fold (P_{NMDG}^{+}/P_{Na}^{+} from 0.05 to 0.22). The dilated state of TRPA1 was able to increase intracellular YO-PRO-1^{+2} when stimulated by high doses of AITC and continued to increase for approximately 50 min [86, 87].

Banke and collaborators reported that the pore formation mechanism is reversible. In this work, the authors aimed to reveal the mechanism underlying the permeabilization of TRPA1, in which the gap junction antagonists carbenoxolone, probenecid, and quin did not stimulate TRPA1 activity or dye uptake in the absence of agonists and had no significant effect on AITC-induced YO-PRO-1 uptake. Furthermore, MAPKs, PKA, or PKC do not induce a response, suggesting a dilation mechanism [88].

Outside-out recordings with a normal external solution demonstrated that the TRPA1 channel had a conductance of ~100 pS at −60 mV in HEK293 cells transfected with mouse TRPA1 cDNA [89]. Compared with Na^{+}, TRPA1 facilitates the uptake of organic cations: monomethylammonium (MA^{+}), DMA^{+}, trimethylammonium ($TriMA^{+}$), TMA^{+}, and $NMDG^{+}$ (**Table A1**). Consistent with the electrophysiological results, TRPA1 activation in CHO cells mediated FM-143^{+2} uptake, suggesting pore diameters of 11 and 13.8 Å [8]. Interestingly, the TRPA1 structure was analyzed by single-particle EM at 16 Å resolution [90].

TRPA1 can facilitate the uptake of drugs such as gentamicin. Studies performed on outer hair cells from young postnatal mice revealed the permeability of the gentamicin–Texas Red conjugate in wild-type animals but not in TRPA1 knockout animals [159]. Furthermore, the lidocaine derivative QX-314 passes through TRPA1 expressed in HEK293 cells after carvacrol stimulation via intracellular cytokine-independent mechanisms [150, 151].

Interestingly, our group proposed that TRPA1 is recruited after P2X7R activation. After $eATP^{-4}$/P2X7R activation, the selective TRPA1 antagonist (HC-030031) reduced fluorescence by approximately 60% for PI^{+2} and EB^{+} uptake but not for anionic dyes. In this same study, in cells that do not express P2X7Rs, TRPA1 emitted currents after stimulation with AITC but not after stimulation with $eATP^{-4}$ or $BzATP^{-4}$, which is responsible for the P2X7R-independent functionality of TRPA1 for ionic functions. Future studies that address these signaling pathways are needed [56].

2.14 Transient receptor potential melastatin 8 (TRPM8)

TRPM8 is a thermosensor for cold temperatures (<20°C) and a cation-permeable menthol-responsive channel [141]. TRPM8 showed relatively low permeability to Ca^{2+} and low selectivity for the monovalent cations P_{Ca}^{2+}/P_{Na}^{+}, P_{K}^{+}/P_{Na}^{+}, and P_{C}^{S+}/P_{K}^{+} (3.3, 1.2 and 1, respectively). Furthermore, the cell-attached configuration showed that transient openings promoted a slope conductance of 83 pS at positive potentials up to 150 mV [91]. Conductance differences may have occurred because of the temperature and stimulus at which the electrophysiological recordings were obtained. For example, 83 pS was measured at 22°C when stimulated by menthol, and 70 pS was measured at 20°C. Furthermore, recordings at a temperature of 30°C allowed a conductance of up to 95 pS for TRPM8 in HEK293 cells [92].

Initially, TRPM8 did not induce YO-PRO-1^{+2} dye in the intracellular compartment. In this study, human TRPM8 was cloned and inserted into the pcDNA3.1/V5-His TOPO vector and transiently expressed in HEK293-F or HeLa cells. The agonists menthol and ilicin activated Ca^{2+} efflux but not large molecule uptake in HEK293-F cells. Likewise, menthol did not evoke $NMDG^{+}$ or Na^{+} currents, demonstrating the absence of permeabilization for this receptor [87].

In contrast, a follow-up study revealed the opposite results: the canine TRPM8 receptor was cloned and inserted into pcDNA4/TO. In a dose-dependent response to the agonist ilicin, CHO-T-REx cells exhibited significant uptake of intracellular YO-PRO-1^{+2}. Interestingly, this study revealed that YO-PRO-1^{+2} uptake occurred via TRPM8 even in the presence of carbenoxolone, a hemichannel blocker, which suggests the absence of second messengers in YO-PRO-1^{+2} uptake and reinforces the dilation mechanism [88]. The conflicting results between these studies may be due to the adopted protocol and experimental conditions.

A few years later, McCoy and colleagues demonstrated that TRPM8 leads to the uptake of large molecules after stimulation by two different agonists. A dose-dependent intracellular influx of PO-PRO-3^{+2}, a variant of YO-PRO-1^{+2}, was induced in HEK293 cells transfected with rat TRPM8 after agonist treatment. PO-PRO-3 uptake by 87.4% of trigeminal sensory neurons was also detected via TRPM8 [16, 18, 93, 94].

There is not much information about the pore size of TRPM8. The TRPM8 structural model demonstrated that the ion pathway forms a wider conduction point of ~9.1 Å in diameter, allowing the passage of hydrated cations [18, 94]. However, the fact that TPRM8 facilitates YO-PRO-1 dye uptake suggests that the pore diameter is >11 Å.

In this sense, a whole-cell configuration with a single voltage step of up to 10 mV (from a holding potential of −80 mV) was used to evaluate whether the intracellular uptake of the compound QX-314 could be mediated via TRPM8. The authors used the mouse neuroblastoma cell line Neuro2A (N2A), which expresses voltage-dependent Na^+ channels, without the presence of TRPM8, and a group of cells with transfected rat TRPM8, making it possible to measure whether stimulation with the TRPM8 agonist (WS -12) and QX-314 generates some effect on Na^+ currents. There was no difference between the control and stimulated cells, whereas in cells expressing TRPM8, the Na^+ current was reduced by 87.6 ± 4.6% after 120 s of exposure to the agonist and QX-314. These findings indicate the real participation of TRPM8 in the uptake of drugs [16].

3. Pore-inducing chemical agents with ion channel activity

3.1 Maitotoxin

Maitotoxin (MTX) is a potent toxin originating from the dinoflagellate *Gambierdiscus toxicus* that causes severe intoxication called ciguatera poisoning. M|TX is an intriguing compound, mainly because of its ionotropic activity and marked permeability to Ca^{2+}. MTX is commonly associated with cytolytic/oncotic pores [95]. MTX increases Ca^{2+} influx in bovine aortic endothelial cells [160], spermatogenic cells [161], GH4C1 rat pituitary cells [162] and human skin fibroblasts [163]. Several studies have suggested that MTX is an agonist of voltage-dependent Ca^{2+} channels. However, MTX stimulation has been shown to generate ionic pores permeable to Na^+ and K^+ in the oocytes and single ventricular cells of guinea pigs [96, 97].

The biophysical properties of MTX-evoked currents in oocytes indicate no selectivity for monovalent cations, with a slight preference for $NH4^+ > K^+ > Na^+ > Li^+$, but low conductance for Ca^{2+} and minimum conductance for Mn^{2+} may predict that the channels stimulated under the experimental conditions in question have low permeability for divalent cations [98]. In terms of the concentration of di- and trivalent cations in the medium, the MTX current decreased in the order of $La^{3+} > Gd^{3+} > Ni^{2+} > Ca^{2+}$. Thus, the ionic response to MTX stimulation depends on the condition or cell type. Cell-attached recordings revealed that the current–voltage relationship of the single-channel conductance was approximately 16 pS in the ventricular myocytes of guinea pigs [97]. Interestingly, three different hyperpolarizing unitary conductances were recorded after MTX application: 103.7 pS in the concentration range < 100 pM and 260.4 and 466.9 pS in the concentration range > 150 pM [96].

The phenomenon promoted by MTX was similar to that promoted by P2X7R, as both allow the passage of Ca^{2+}, Na^+, and K^+ in addition to the capture of fluorescent dyes and a high conductance of up to 400 pS. Whole-cell currents in human fibroblasts allowed the replacement of the Na^+ bath with $NMDG^+$. Furthermore, EB^+, YO-PRO-1^{+2}, POPO-3^{+4}, and fura-2^{-5} uptake suggested the progressive expansion of membrane pores [99]. However, whether these phenomena are coincidental or whether MTX functions as a P2X7R activator remains unclear.

Interestingly, this same group reported that the upregulation of P2X7R expression in THP-1 cells after 48 h of LPS and IFN-γ incubation reduced the number of pores induced by MTX. However, the permeability of both pores was similar for permeating EB^+, YO-PRO-1^{+2}, and POPO-3^{+4} [17]. Furthermore, when P2X7R was irreversibly inhibited by oxidized ATP, MTX pore activity was unaffected [100]. These findings suggest that both pore types present similar characteristics with distinct intracellular pathways for activating these pores [21]. According to the probes, the pore size may be >14 Å.

In this context, TRPs may interact with the MTX pore. TRP channels are nonselective and highly Ca^{2+}-permeable. However, some previously demonstrated characteristics of MTX in oocytes do not correspond to the conductance of some TRPs, even those known to form pores (10–100 pS) [141, 164]. Some authors associate MTX with the canonical channel 1 of the transient receptor potential (TRPC1) [165, 166]. Although there are similarities, such as conductance (TRPC1 in HEK293 cells ~17 pS) and (MTX in guinea pig myocytes ~16 pS), Ca^{2+} permeability and inhibition by trivalent cations such as Gd^{3+} and La^{3+} [97, 167], the formation of large pores of TPRC1 has not been well elucidated, whereas the pores induced by MTX exhibit high conductance (> 400 pS) in oocyte electrophysiological recordings [96, 99]. New studies in this regard are necessary.

3.2 Ionomycin

Ionomycin is an antibiotic produced by the bacterium *Streptomyces conglobatus* and is also a Ca^{2+} ionophore [168, 169]. Initially, ionomycin was reported to induce exocytotic fusion pores, with an ionic conductance of 150 pS, which decreased to 35 pS [170]. Our group observed second messengers in the formation of P2X7R pores in 2BH4 cells, and ionomycin, which was applied at low concentrations, was used to potentiate eATP-4. Thus, Ca^{2+} was an essential factor in the P2X7R pores. Interestingly, stimulation with ionomycin in a cell-attached configuration at a concentration of 10 μM without $eATP^{-4}$ induced EB^{+} uptake and conductance of 412 pS, values slightly close to those acquired by P2X7R (445 pS) [49].

Another Ca^{2+} ionophore (A – 23,187) and thapsigargin (an endoplasmic reticulum Ca^{2+} ATPase blocker) also mediate EB^{+} uptake but have slower kinetics than ionomycin does [49]. In another study focused on this subject, the ionic conductance properties of ionomycin in 2BH4, IT45-RI, and HEK293 cells were 421, 411, and 415 pS, respectively. In 2BH4 cells, ionomycin induced EB^{+}, PI^{+2}, and LY^{-2} uptake but not trypan blue or TRITC-dextran uptake, suggesting a pore size limit of ~900 Da. According to the probes used, the pore size could be estimated to be up to 14 Å [101].

The actual mechanism by which Ca^{2+} affects pore opening is unknown. The CAHLM1 receptor has high selectivity for Ca^{2+} and is large enough to cross through membrane high-molecular-weight cations and anions. However, CAHLM1 conductance (24 pS) did not correspond to that found after stimulation with ionomycin (421 pS). Furthermore, the data point to the influence of intracellular pathways as possible promoters of ionomycin-associated pores, such as reduced levels of available Ca^{2+}, phospholipase C, the p38 MAPK pathway, and the cytoskeleton rearrangement pathway, but not P2X7R-dependent pathways, hemichannels (connexins and pannexins), camp-activated kinases, and PKC [101].

4. Perspectives in the study of pore-forming channels

The activation of the ion channel followed by the opening of the high conductance pore is associated with processes such as pain, inflammation, and cell death, mainly involving the P2X7 and TRPV1 receptors, which stand out with greater scope. Some of these receptors lack information on their diameter and permeabilization limit (**Table 1**) and on the presence of second messengers and cell type specificity; further studies of electrophysiology with specific configurations would be interesting [3, 4].

Another point of importance is the possible molecular interactions that occur between amino acids lining the pore and small ions or molecules of greater molecular weight. The study of fluorescent dyes provides a good overview of pore diameter in studies of these channels in combination with electrophysiology. However, most of the dyes used in these studies have mono- and divalent cationic or anionic characteristics, with few explorations with higher charges, mainly with fluorophores with charges of +3 and + 4. The charge relationship along the passage through the channel and interaction with amino acid residues could be an informative complement to the physicochemical study of these channels [3].

Interestingly, the use of pore-forming channels to increase the permeability of membrane-impermeable drugs could be a reliable therapeutic tool. In 2015, it was proposed that the activation of P2X7R with $eATP^{-4}$ and the induction of pore opening could facilitate entry into the cytoplasm of antibacterial agents against *Mycobacterium tuberculosis*. In this case, cells infected by the parasite benefit synergistically from activating P2X7R, facilitating the entry of drugs such as rifampicin (822.94 Da), ethambutol (277.2 Da) and ethionamide (166.2 Da) into cells [171]. P2X7Rs act synergistically with drugs, such as methylene blue and doxorubicin (both with a charge of +1). Similarly, TRP channels can increase the activity of drugs such as the +1 charged lidocaine derivative QX-314 and the antibiotic gentamicin (**Table A3**) [148, 150]. Thus, enhancing the action of impermeable drugs may be possible.

5. Conclusion

Pore-forming channels are related to cellular electrochemical functionality and pathophysiological stimuli. Patch clamp techniques combined with fluorescence marked a milestone in the history and study of pore-forming channels. Through both methods, it was possible to calculate and estimate the ionic conductance, cutoff, and pore size parameters. More prominently, we identified the hemichannel receptors Cx, MAC, P2X7R, Panx-1, Pl-VDAC1, TPRV1, and TRPA1, which are capable of permeating electrically charged molecules with molecular weights ranging from 900 to 1000 Da. Thus, strategies involving these channels as facilitators of the permeability of membrane-impermeable drugs are suggested as promising tools for treating diseases.

Acknowledgements

The fellowships granted by CNPq (316568/2021-0) and CAPES – Finance Code 001 and FAPERJ (E-26/203.246/2017, E-26/203.146/2017, E-26/211.025/2019, E-26/200.982/2021, and E-26/201.369/2021) are gratefully acknowledged. We are grateful to LAPSA – IOC, FIOCRUZ, Brazil, and PBV – UFRJ, Brazil.

Conflict of interest

The authors declare that they have no conflicts of interest.

Appendix

See **Table A1-A3**.

Organic cations	**Molecular weigh (Da)**	**Molecular dimension (Å)**	**Charge**	**Reference**
2-(methyl-amino)-ethanol) (2-MAE)	75.11	6.2	+1	[7]
Dimethylammonium (DMA)	46.091	4.6	+1	[8]
N-methyl-d-glucamine (NMDG)	195.21	9	+1	[8]
Monomethylammonium	32.065	3.6	+1	[8]
Trimethylammonium (TriMA)	60.12	5.2	+1	[8]
Tetramethylammonium (TMA)	74,14	5.8	+1	[8]
Tetraethylammonium (TEA)	130,25	8	+1	[9]
Tetra-butylammonium (TBA)	242,470	9.88	+1	[9]
Tris(hidroximetil)aminometano (Tris)	121.14	7.4	+1	[7]
Organic anions	**Molecular weigh (Da)**	**Molecular dimension (Å)**	**Charge**	**Reference**
Acetate	59.044	4.84	-1	[10]
Adenosine triphosphate (ATP)	507,18	13	-4	[10]
Adenosine diphosphate (ADP)	424.18	N/A	-2	[11]
Adenosine monophosphate (AMP)	347.22	N/A	-2	[11]
Formate	46,03	4.24	-1	[10]
Gluconate	196,16	6.98	-1	[10]
Glucoheptonate	248.16	7.86	-1	[10]
Glucuronate	193.13	7.26	-1	[10]
Glutamate	146.12	6.9	-1	[10]
Lactobionate	357.29	9.74	-1	[10]
Methanesulfonate	95.10	5.34	-1	[10]
Propanoate	73.071	5.34	-1	[10]
Pyruvate	87.05	5.42	-1	[10]
MgATP	529.47	12	-2	[10]
Polyethylene glycol	**Molecular weigh (Da)**	**Molecular dimension (Å)**	**Charge**	**Reference**
Ethylene glycol	62	4.4	Neutral	[12]
Diethylene glycol	106	5.8	Neutral	[12]
Triethylene Glycol	150	6.8	Neutral	[12]
Glycerol	92.09	6.2	Neutral	[10]
PEG 200	200	8	Neutral	[12]
PEG 300	300	9.6	Neutral	[12]
PEG 400	400	11.2	Neutral	[12]
PEG 600	600	13.8	Neutral	[12]
PEG 1000	1000	17.8	Neutral	[12]
PEG 1540	1540	~23	Neutral	[12]
PEG 2000	2000	~28	Neutral	[10]
PEG 3500	3500	> 30	Neutral	[12]
PEG 4000	4000	~40	Neutral	[10]

Table A1.
Organic molecules of cations and anions and polyethylene glycol are used to study pore-forming ion channels. The size of the compounds is based on the molecular diameter. The work that provided information about the radius was standardized to the diameter at twice the value.

Cationic fluorophore	Molecular weigh (Da)	Molecular dimension (Å)	Charge	Reference
DAPI	277.32	N/A	+1	[13]
Ethidium bromide	394.3	10.3	+1	[14]
FM1–43	451.7	10.5	+2	[15]
PO-PRO-3	351.5	N/A	+2	[16]
POPO-3	715.0	N/A	+4	[17]
Propidium iodide	668	13.8	+2	[18]
TOTO-1	795.2	N/A	+4	[19]
YO-PRO-1	629.3	10.6	+2	[18]
Anionic fluorophore	**Molecular weigh (Da)**	**Molecular dimension (Å)**	**Charge**	**Reference**
Carboxyfluorescein	376	8.2	-1	[14]
Alexa350	350	5.2	-1	[9]
Alexa488	570	10.5	-2	[9]
Alexa594	760	13.8	-2	[9]
Alexa633	>1000	N/A	-2	[9]
Calcein	622.5	13	-4	[18]
Evans blue	960.8	N/A	-4	[20]
Fluorescein	332.3	N/A	-1	[21]
Fluorescein isothiocyanate	389.4	> 10	-1	[22]
Fura-2	831	N/A	-5	[20]
LuciferYellow	443	9.9	-2	[14]
Trypan blue	960.81	N/A	-4	[20]

Table A2.
Fluorescent dyes used in studies of pore-forming ion channels. The size of the compounds is based on the molecular diameter. The work that provided information about the radius was standardized to the diameter at twice the value.

Ion channel	Drugs	Reference
P2X7R	Methylene blue Doxorubicin	[116, 117]
TRPV1	Qx314 Gentamicin Gabapentin	[147–151]
TRPV2	Temozolomide Carmustine Doxorubicin Bortezomib	[155, 156]
TRPV4	Gentamicin	[148, 157]
TRPA1	Qx314 Gentamicin	[150, 151, 159]
TRPM8	Qx314	[94]

Table A3.
Drugs permeable to large conductance channels.

Author details

Guilherme Pegas Teixeira[1,3], Leandro Rocha[2,3] and Robson Xavier Faria[1*]

1 Laboratory for Environmental Health Assessment and Promotion, Oswaldo Cruz Institute, Manguinhos, Rio de Janeiro, Brazil

2 Laboratory of Natural Products Technology, Faculty of Pharmacy, Fluminense Federal University, Santa Rosa, Niterói, Brazil

3 Vegetal Biotechnology and Bioprocesses, Rio de Janeiro Federal University, Rio de Janeiro, Brazil

*Address all correspondence to: robson.xavier@gmail.com

References

[1] Zimmerberg J. Membrane biophysics. Current Biology. 2006;**16**:272-276. DOI: 10.1016/j.cub.2006.03.050

[2] Lawrence JH, Tomaselli GF, Marban E. Ion channels: Structure and function. Heart Disease and Stroke. 1993;**2**:75-80. Available from: http://europepmc.org/abstract/MED/7511968

[3] Teixeira G, Faria R. Pore-forming proteins: Fluorescent dyes to study the channel functionality and biophysical properties. In: Grigoryeva N, editor. Fluorescence Methods for Investigation of Living Cells and Microorganisms. Rijeka: IntechOpen; 2020. pp. 1-15. DOI: 10.5772/intechopen.92870

[4] Ferreira L, Pereira L, Faria R. Fluorescent dyes as a reliable tool in P2X7 receptor-associated pore studies. Journal of Bioenergetics and Biomembranes. 2015;**47**:1-25. DOI: 10.1007/s10863-015-9613-0

[5] Kamiya K, Osaki T, Nakao K, Kawano R, Fujii S, Misawa N, et al. Electrophysiological measurement of ion channels on plasma/organelle membranes using an on-chip lipid bilayer system. Scientific Reports. 2018;**8**:1-9. DOI: 10.1038/s41598-018-35316-4

[6] Hill CL, Stephens GJ. An introduction to patch clamp recording BT—Patch clamp electrophysiology: Methods and protocols. In: Dallas M, Bell D, editors. Patch Clamp Electrophysiology. New York, NY: Springer US; 2021. pp. 1-19. DOI: 10.1007/978-1-0716-0818-0_1

[7] Eickhorst AN, Berson A, Cockayne D, Lester HA, Khakh BS. Control of $P2X_2$ channel permeability by the cytosolic domain. The Journal of General Physiology. 2002;**120**:119-131. DOI: 10.1085/jgp.20028535

[8] Karashima Y, Prenen J, Talavera K, Janssens A, Voets T, Nilius B. Agonist-induced changes in Ca^{2+} permeation through the nociceptor Cation Channel TRPA1. Biophysical Journal. 2010;**98**:773-783. DOI: 10.1016/j.bpj.2009.11.007

[9] Siebert A, Ma Z, Grevet J, Demuro A, Parker I, Foskett J. Structural and functional similarities of calcium homeostasis modulator 1 (CALHM1) ion channel with connexins, pannexins and innexins. The Journal of Biological Chemistry. 2013;**288**:6140-6153. DOI: 10.1073/pnas.1204023109

[10] Sabirov RZ, Okada Y. Wide nanoscopic pore of maxi-anion channel suits its function as an ATP-conductive pathway. Biophysical Journal. 2004;**87**:1672-1685. DOI: 10.1529/biophysj.104.043174

[11] Harris AL. Connexin channel permeability to cytoplasmic molecules. Progress in Biophysics and Molecular Biology. 2007;**94**:120-143. DOI: 10.1016/j.pbiomolbio.2007.03.011

[12] Gong X-Q, Nicholson BJ. Size selectivity between gap junction channels composed of different connexins. Cell Communication & Adhesion. 2001;**8**:187-192. DOI: 10.3109/15419060109080721

[13] Gaete PS, Kumar D, Fernandez CI, Valdez-Capuccino JM, Bhatt A, Jiang W, et al. Large-pore connexin hemichannels function as molecule transporters independently of ion conduction. bioRxiv. 2024; PRE-PRINT. Available from: http://biorxiv.org/content/early/2024/02/20/2024.02.20.581300.abstract

[14] Kanaporis G, Brink PR, Valiunas V. Gap junction permeability: Selectivity for anionic and cationic probes. American Physiological Society Journal. 2010;**300**:600-609. DOI: 10.1152/ajpcell.00316.2010

[15] Meyers JR, MacDonald RB, Duggan A, Lenzi D, Standaert DG, Corwin JT, et al. Lighting up the senses: FM1-43 loading of sensory cells through nonselective ion channels. The Journal of Neuroscience. 2003;**23**:4054-4065. DOI: 10.1523/JNEUROSCI.23-10-04054.2003

[16] McCoy DD, Palkar R, Yang Y, Ongun S, McKemy DD. Cellular permeation of large molecules mediated by TRPM8 channels. Neuroscience Letters. 2017;**639**:59-67. DOI: 10.1016/j.neulet.2016.12.063

[17] Schilling WP, Wasylyna T, Dubyak GR, Humphreys BD, Sinkins WG. Maitotoxin and P2Z/$P2X_7$purinergic receptor stimulation activate a common cytolytic pore. American Physiological Society Journal. 1999;**277**:766-776. DOI: 10.1152/ajpcell.1999.277.4.C766

[18] Sözer EB, Pocetti CF, Vernier PT. Transport of charged small molecules after electropermeabilization—Drift and diffusion. BMC Biophysics. 2018;**11**:1-11. DOI: 10.1186/s13628-018-0044-2

[19] Cankurtaran-Sayar S, Sayar K, Ugur M. $P2X_7$ receptor activates multiple selective dye-permeation pathways in RAW 264.7 and human embryonic kidney 293 cells. Molecular Pharmacology. 2009;**76**:1323-1332. DOI: 10.1124/mol.109.059923

[20] Nuttle LC, El-Moatassim C, Dubyak GR. Expression of the pore-forming P2Z purinoreceptor in *Xenopus oocytes* injected with poly(a)+ RNA from murine macrophages. Molecular Pharmacology. 1993;**44**:93-101

[21] Ferreira LGB, Reis RAM, Alves LA, Faria RX. Intracellular signaling pathways integrating the pore associated with $P2X_7R$ receptor with other large pores. In: Kaneez FS, editor. Patch Clamp Technique. Rijeka: IntechOpen; 2012. pp. 1-20. DOI: 10.5772/38842

[22] Browne LE, Compan V, Bragg L, North RA. $P2X_7$ receptor channels allow direct permeation of nanometer-sized dyes. The Journal of Neuroscience. 2013;**33**:3557-3566. DOI: 10.1523/JNEUROSCI.2235-12.2013

[23] Ma Z, Tanis JE, Taruno A, Foskett JK. Calcium homeostasis modulator (CALHM) ion channels. Pflügers Archiv: European Journal of Physiology. 2016;**468**:395-403. DOI: 10.1007/s00424-015-1757-6

[24] Vingtdeux V, Chandakkar P, Zhao H, Blanc L, Ruiz S, Marambaud P. CALHM1 ion channel elicits amyloid-β clearance by insulin-degrading enzyme in cell lines and in vivo in the mouse brain. Journal of Cell Science. 2015;**128**:2330-2338. DOI: 10.1242/jcs.167270

[25] Ma Z, Siebert AP, Cheung K-H, Lee RJ, Johnson B, Cohen AS, et al. Calcium homeostasis modulator 1 (CALHM1) is the pore-forming subunit of an ion channel that mediates extracellular Ca^{2+} regulation of neuronal excitability. Proceedings of the National Academy of Sciences. 2012;**109**:1963-1971. DOI: 10.1073/pnas.1204023109

[26] Taruno A, Vingtdeux V, Ohmoto M, Ma Z, Dvoryanchikov G, Li A, et al. CALHM1 ion channel mediates purinergic neurotransmission of sweet, bitter and umami tastes. Nature. 2013;**495**:223-226. DOI: 10.1038/nature11906

[27] Valiunas V. Biophysical properties of connexin-45 gap junction hemichannels studied in vertebrate cells. The Journal of General Physiology. 2002;**119**:147-164. DOI: 10.1085/jgp.119.2.147

[28] Beyer E, Steinberg T. Evidence that the gap junction protein connexin-43 is the ATP-induced pore of mouse macrophages. The Journal of Biological Chemistry. 1991;**266**:7971-7974

[29] Musa H, Veenstra RD. Voltage-dependent blockade of connexin40 gap junctions by spermine. Biophysical Journal. 2003;**84**:205-219. DOI: 10.1016/S0006-3495(03)74843-7

[30] Sánchez HA, Orellana JA, Verselis VK, Sáez JC. Metabolic inhibition increases activity of connexin-32 hemichannels permeable to Ca^{2+} in transfected HeLa cells. American Physiological Society Journal. 2009;**297**:665-678. DOI: 10.1152/ajpcell.00200.2009

[31] Goldberg GS, Valiunas V, Brink PR. Selective permeability of gap junction channels. Biochimica et Biophysica Acta - Biomembranes. 2004;**1662**:96-101. DOI: 10.1016/j.bbamem.2003.11.022

[32] Sáez JC, Connor JA, Spray DC, Bennett MV. Hepatocyte gap junctions are permeable to the second messenger, inositol 1,4,5-trisphosphate, and to calcium ions. Proceedings of the National Academy of Sciences. 1989;**86**:2708-2712. DOI: 10.1073/pnas.86.8.2708

[33] Kanaporis G, Mese G, Valiuniene L, White TW, Brink PR, Valiunas V. Gap junction channels exhibit connexin-specific permeability to cyclic nucleotides. The Journal of General Physiology. 2008;**131**:293-305. DOI: 10.1085/jgp.200709934

[34] Wang H-Z, Veenstra RD. Monovalent ion selectivity sequences of the rat connexin43 gap Junction channel. The Journal of General Physiology. 1997;**109**:491-507. DOI: 10.1085/jgp.109.4.491

[35] Kang J, Kang N, Lovatt D, Torres A, Zhao Z, Lin J, et al. Connexin 43 hemichannels are permeable to ATP. The Journal of Neuroscience. 2008;**28**:4702-4711. DOI: 10.1523/JNEUROSCI.5048-07.2008

[36] Beblo DA, Veenstra RD. Monovalent cation permeation through the connexin40 gap Junction channel: Cs, Rb, K, Na, Li, TEA, TMA, TBA, and effects of anions Br, Cl, F, acetate, aspartate, glutamate, and NO_3. The Journal of General Physiology. 1997;**109**:509-522. DOI: 10.1085/jgp.109.4.509

[37] Weber PA, Chang H-C, Spaeth KE, Nitsche JM, Nicholson BJ. The permeability of gap junction channels to probes of different size is dependent on connexin composition and permeant-pore affinities. Biophysical Journal. 2004;**87**:958-973. DOI: 10.1529/biophysj.103.036350

[38] Veenstra RD, Wang HZ, Beyer EC, Brink PR. Selective dye and ionic permeability of gap junction channels formed by connexin45. Circulation Research American Heart Association. 1994;**75**:483-490. DOI: 10.1161/01.RES.75.3.483

[39] Burnstock G, Verkhratsky A. Receptors for purines and pyrimidines BT—Purinergic signalling and the nervous system. In: Burnstock G, Verkhratsky A, editors. Berlin, Heidelberg: Springer; 2012. pp. 119-244. DOI: 10.1007/978-3-642-28863-0_5

[40] Ding S, Sachs F. Single Channel properties of $P2X_2$ purinoceptors.

The Journal of General Physiology. 1999;**113**:695-720. DOI: 10.1085/jgp.113.5.695

[41] Virginio C, MacKenzie A, Rassendren FA, North RA, Surprenant A. Pore dilation of neuronal P2X receptor channels. Nature Neuroscience. 1999;**2**:315-321. DOI: 10.1038/7225

[42] Mio K, Ogura T, Yamamoto T, Hiroaki Y, Fujiyoshi Y, Kubo Y, et al. Reconstruction of the $P2X_2$ receptor reveals a vase-shaped structure with lateral tunnels above the membrane. Structure. 2009;**17**:266-275. DOI: 10.1016/j.str.2008.12.007

[43] Chaumont S, Khakh BS. Patch–clamp coordinated spectroscopy shows $P2X_2$ receptor permeability dynamics require cytosolic domain rearrangements but not Panx-1 channels. Proceedings of the National Academy of Sciences. 2008;**105**:12063-12068. DOI: 10.1073/pnas.0803008105

[44] Guimaraes MZP. Isoform specificity of $P2X_2$ purinergic receptor C-terminus binding to tubulin. Neurochemistry International. 2008;**52**:314-320. DOI: 10.1016/j.neuint.2007.06.032

[45] Marques-da-Silva C, Chaves MM, Castro NG, Coutinho-Silva R, Guimaraes MZP. Colchicine inhibits cationic dye uptake induced by ATP in $P2X_2$ and $P2X_7$ receptor-expressing cells: Implications for its therapeutic action. British Journal of Pharmacology. 2011;**163**:912-926. DOI: 10.1111/j.1476-5381.2011.01254.x

[46] Negulyaev YA, Markwardt F. Block by extracellular Mg^{2+} of single human purinergic $P2X_4$ receptor channels expressed in human embryonic kidney cells. Neuroscience Letters. 2000;**279**:165-168. DOI: 10.1016/S0304-3940(99)00976-3

[47] Khakh BS, Bao XR, Labarca C, Lester HA. Neuronal P2X transmitter-gated cation channels change their ion selectivity in seconds. Nature Neuroscience. 1999;**2**:322-330. DOI: 10.1038/7233

[48] Latapiat V, Rodríguez-Tirado F, Godoy F, Montenegro F, Barrera N, Huidobro-Toro J. $P2X_4$ receptor *in silico* and electrophysiological approaches reveal insights of ivermectin and zinc allosteric modulation. Frontiers in Pharmacology. 2017 Dec;**15**(8):1-13. DOI: 10.3389/fphar.2017.00918

[49] Faria RX, DeFarias FP, Alves LA. Are second messengers crucial for opening the pore associated with $P2X_7$ receptor? American Physiological Society Journal. 2005;**288**:260-271. DOI: 10.1152/ajpcell.00215.2004

[50] Pelegrín P. Many ways to dilate the $P2X_7$ receptor pore. British Journal of Pharmacology. 2011;**163**:908-911. DOI: 10.1111/j.1476-5381.2011.01325.x

[51] Riedel T, Lozinsky I, Schmalzing G, Markwardt F. Kinetics of $P2X_7$ receptor-operated single channels currents. Biophysical Journal. 2007;**92**:2377-2391. DOI: 10.1529/biophysj.106.091413

[52] Chessell IP, Grahames CBA, Michel AD, Humphrey PPA. Dynamics of $P2X_7$ receptor pore dilation: Pharmacological and functional consequences. Drug Development Research. 2001;**53**:60-65. DOI: 10.1002/ddr.1171

[53] Sun C, Heid ME, Keyel PA, Salter RD. The second transmembrane domain of $P2X_7$ contributes to dilated pore formation. PLoS One. 2013;**8**:1-10. DOI: 10.1371/journal.pone.0061886

[54] Jiang L-H, Rassendren F, Mackenzie A, Zhang Y-H, Surprenant A, North RA. N-methyl-d-glucamine and propidium dyes utilize different permeation pathways at rat P2X$_7$ receptors. American Physiological Society Journal. 2005;**289**:1295-1302. DOI: 10.1152/ajpcell.00253.2005

[55] Coutinho-Silva R, Persechini PM. P2Z purinoceptor-associated pores induced by extracellular ATP in macrophages and J774 cells. American Physiological Society Journal. 2005;**273**:1793-1800. DOI: 10.1152/ajpcell.1997.273.6.C1793

[56] Faria RX, Reis RAM, Ferreira LGB, Cezar-de-Mello PFT, Moraes MO. P2X$_7$R large pore is partially blocked by pore forming proteins antagonists in astrocytes. Journal of Bioenergetics and Biomembranes. 2016;**48**:309-324. DOI: 10.1007/s10863-016-9649-9

[57] Ma W, Compan V, Zheng W, Martin E, North RA, Verkhratsky A, et al. Pannexin 1 forms an anion-selective channel. Pflügers Archiv: European Journal of Physiology. 2012;**463**:585-592. DOI: 10.1007/s00424-012-1077-z

[58] Bao L, Locovei S, Dahl G. Pannexin membrane channels are mechanosensitive conduits for ATP. FEBS Letters. 2004;**572**:65-68. DOI: 10.1016/j.febslet.2004.07.009

[59] Dubyak GR. Both sides now: Multiple interactions of ATP with pannexin-1 hemichannels. Focus on "a permeant regulating its permeation pore: Inhibition of pannexin 1 channels by ATP". American Physiological Society Journal. 2009;**296**:235-241. DOI: 10.1152/ajpcell.00639.2008

[60] Kuzuya M, Hirano H, Hayashida K, Watanabe M, Kobayashi K, Terada T, et al. Structures of human pannexin-1 in nanodiscs reveal gating mediated by dynamic movement of the N-terminus and phospholipids. Science Signaling. 2024;**15**:1-10. DOI: 10.1126/scisignal.abg6941

[61] Sahu G, Sukumaran S, Bera AK. Pannexins form gap junctions with electrophysiological and pharmacological properties distinct from connexins. Scientific Reports. 2014;**4**:1-9. DOI: 10.1038/srep04955

[62] Sabirov RZ, Merzlyak PG. Plasmalemmal VDAC controversies and maxi-anion channel puzzle. Biochimica et Biophysica Acta - Biomembrane. 2012;**1818**:1570-1580. DOI: 10.1016/j.bbamem.2011.09.024

[63] Guibert B, Dermietzel R, Siemen D. Large conductance channel in plasma membranes of astrocytic cells is functionally related to mitochondrial VDAC-channels. The International Journal of Biochemistry & Cell Biology. 1998;**30**:379-391. DOI: 10.1016/S1357-2725(97)00137-4

[64] Mannella CA, Forte M, Colombini M. Toward the molecular structure of the mitochondrial channel, VDAC. Journal of Bioenergetics and Biomembranes. 1992;**24**:7-19. DOI: 10.1007/BF00769525

[65] Ujwal R, Cascio D, Colletier J-P, Faham S, Zhang J, Toro L, et al. The crystal structure of mouse VDAC1 at 2.3 Å resolution reveals mechanistic insights into metabolite gating. Proceedings of the National Academy of Sciences. 2008;**105**:17742-17747. DOI: 10.1073/pnas.0809634105

[66] Sabirov RZ, Dutta AK, Okada Y. Volume-dependent Atp-conductive large-conductance anion channel as a pathway for swelling-induced Atp release. The Journal of General

Physiology. 2001;**118**:251-266. DOI: 10.1085/jgp.118.3.251

[67] Sabirov RZ, Okada Y. The maxi-anion channel: A classical channel playing novel roles through an unidentified molecular entity. The Journal of Physiological Sciences. 2009;**59**:3-21. DOI: 10.1007/s12576-008-0008-4

[68] Dutta A, Sabirov R, Uramoto H, Okada Y. Role of ATP-conductive anion channel in ATP release from neonatal rat cardiomyocytes in ischemic or hypoxic conditions. The Journal of Physiology. 2004;**559**:799-812. DOI: 10.1113/jphysiol.2004.069245

[69] Bautista D, Julius D. Fire in the hole: Pore dilation of the capsaicin receptor TRPV1. Nature Neuroscience. 2008;**11**:528-529. DOI: 10.1038/nn0508-528

[70] Premkumar LS, Agarwal S, Steffen D. Single-channel properties of native and cloned rat vanilloid receptors. The Journal of Physiology. 2002;**545**:107-117. DOI: 10.1113/jphysiol.2002.016352

[71] Chung M-K, Güler AD, Caterina MJ. TRPV1 shows dynamic ionic selectivity during agonist stimulation. Nature Neuroscience. 2008;**11**:555-564. DOI: 10.1038/nn.2102

[72] Moiseenkova-Bell VY, Stanciu LA, Serysheva II, Tobe BJ, Wensel TG. Structure of TRPV1 channel revealed by electron cryomicroscopy. Proceedings of the National Academy of Sciences. 2008;**105**:7451-7455. DOI: 10.1073/pnas.0711835105

[73] Jansson ET, Trkulja CL, Ahemaiti A, Millingen M, Jeffries GDM, Jardemark K, et al. Effect of cholesterol depletion on the pore dilation of TRPV1. Molecular Pain. 2013;**9**:1741-1744. DOI: 10.1186/1744-8069-9-1

[74] Zhang F, Hanson SM, Jara-Oseguera A, Krepkiy D, Bae C, Pearce LV, et al. Engineering vanilloid-sensitivity into the rat TRPV2 channel. eLife. 2016;**5**:e16409. DOI: 10.7554/eLife.16409

[75] Egée S, Kaestner L. The transient receptor potential vanilloid type 2 (TRPV2) channel–a new druggable Ca^{2+} pathway in red cells, implications for red cell ion homeostasis. Frontiers in Physiology. 2021;**12**:1-5. DOI: 10.3389/fphys.2021.677573

[76] Compan V, Baroja-Mazo A, López-Castejón G, Gomez AI, Martínez CM, Angosto D, et al. Cell volume regulation modulates NLRP3 inflammasome activation. Immunity. 2012;**37**:487-500. DOI: 10.1016/j.immuni.2012.06.013

[77] Zubcevic L, Le S, Yang H, Lee S-Y. Conformational plasticity in the selectivity filter of the TRPV2 ion channel. Nature Structural & Molecular Biology. 2018;**25**:405-415. DOI: 10.1038/s41594-018-0059-z

[78] Schrapers KT, Sponder G, Liebe F, Liebe H, Stumpff F. The bovine TRPV3 as a pathway for the uptake of Na^{+}, Ca^{2+}, and NH^{4+}. PLoS One. 2018;**13**:1-28. DOI: 10.1371/journal.pone.0193519

[79] Chung M-K, Güler AD, Caterina MJ. Biphasic currents evoked by chemical or thermal activation of the heat-gated ion channel, TRPV3. The Journal of Biological Chemistry. 2005;**280**:15928-15941. DOI: 10.1074/jbc.M500596200

[80] Chung M-K, Lee H, Mizuno A, Suzuki M, Caterina MJ. 2-Aminoethoxydiphenyl borate activates and sensitizes the heat-gated ion channel TRPV3. The Journal of Neuroscience. 2004;**24**:5177-5182. DOI: 10.1523/JNEUROSCI.0934-04.2004

[81] Lansky S, Betancourt JM, Zhang J, Jiang Y, Kim ED, Paknejad N, et al. A pentameric TRPV3 channel with a dilated pore. Nature. 2023;**621**:206-214. DOI: 10.1038/s41586-023-06470-1

[82] Voets T, Prenen J, Vriens J, Watanabe H, Janssens A, Wissenbach U, et al. Molecular determinants of permeation through the cation channel TRPV4. The Journal of Biological Chemistry. 2002;**277**:33704-33710. DOI: 10.1074/jbc.M204828200

[83] Watanabe H, Vriens J, Suh SH, Benham CD, Droogmans G, Nilius B. Heat-evoked activation of TRPV4 channels in a HEK293 cell expression system and in native mouse aorta endothelial cells. The Journal of Biological Chemistry. 2002;**277**:47044-47051. DOI: 10.1074/jbc.M208277200

[84] Liedtke W, Choe Y, Martí-Renom MA, Bell AM, Denis CS, AndrejŠali, et al. Vanilloid receptor-related osmotically activated channel (VR-OAC), a candidate vertebrate osmoreceptor. Cell. 2000;**103**:525-535. DOI: 10.1016/S0092-8674(00)00143-4

[85] Nadezhdin KD, Talyzina IA, Parthasarathy A, Neuberger A, Zhang DX, Sobolevsky AI. Structure of human TRPV4 in complex with GTPase RhoA. Nature Communications. 2023;**14**:1-11. DOI: 10.1038/s41467-023-39346-z

[86] Meents JE, Ciotu CI, Fischer MJM. TRPA1: A molecular view. Journal of Neurophysiology. 2018;**121**:427-443. DOI: 10.1152/jn.00524.2018

[87] Chen J, Kim D, Bianchi BR, Cavanaugh EJ, Faltynek CR, Kym PR, et al. Pore dilation occurs in TRPA1 but not in TRPM8 channels. Molecular Pain. 2009;**5**:1743-1744. DOI: 10.1186/1744-8069-5-3

[88] Banke TG, Chaplan SR, Wickenden AD. Dynamic changes in the TRPA1 selectivity filter lead to progressive but reversible pore dilation. American Physiological Society Journal. 2010;**298**:1457-1468. DOI: 10.1152/ajpcell.00489.2009

[89] Nagata K, Duggan A, Kumar G, García-Añoveros J. Nociceptor and hair cell transducer properties of TRPA1, a channel for pain and hearing. The Journal of Neuroscience. 2005;**25**:4052-4061. DOI: 10.1523/JNEUROSCI.0013-05.2005

[90] Cvetkov TL, Huynh KW, Cohen MR, Moiseenkova-Bell VY. Molecular architecture and subunit organization of TRPA1 Ion Channel revealed by electron microscopy*. The Journal of Biological Chemistry. 2011;**286**:38168-38176. DOI: 10.1074/jbc.M111.288993

[91] McKemy DD, Neuhausser WM, Julius D. Identification of a cold receptor reveals a general role for TRP channels in thermosensation. Nature. 2002;**416**:52-58. DOI: 10.1038/nature719

[92] Fernández JA, Skryma R, Bidaux G, Magleby KL, Scholfield CN, McGeown JG, et al. Voltage- and cold-dependent gating of single TRPM8 ion channels. The Journal of General Physiology. 2011;**137**(2):173-195. DOI: 10.1085/jgp.201010498

[93] Yin Y, Zhang F, Feng S, Butay KJ, Borgnia MJ, Im W, et al. Activation mechanism of the mouse cold-sensing TRPM8 channel by cooling agonist and PIP2. Science (80-). 2024;**378**:1-33. DOI: 10.1126/science.add1268

[94] Diver MM, Cheng Y, Julius D. Structural insights into TRPM8 inhibition and desensitization. Science (80-). 2019;**365**:1434-1440. DOI: 10.1126/science.aax6672

[95] Sladeczek F, Schmidt BH, Alonso R, Vian L, Tep A, Yasumoto T, et al. New insights into maitotoxin action. European Journal of Biochemistry. 1988;**174**:663-670. DOI: 10.1111/j.1432-1033.1988.tb14149.x

[96] Egido W, Castrejón V, Antón B, Martínez M. Maitotoxin induces two dose-dependent conductances in *Xenopus oocytes*. Comparison with nystatin effects as a pore inductor. Toxicon. 2008;**51**:797-812. DOI: 10.1016/j.toxicon.2007.12.013

[97] Nishio M, Muramatsu I, Yasumoto T. Na^+-permeable channels induced by maitotoxin in Guinea-pig single ventricular cells. European Journal of Pharmacology. 1996;**297**:293-298. DOI: 10.1016/0014-2999(95)00751-2

[98] Bielfeld-Ackermann A, Range C, Korbmacher C. Maitotoxin (MTX) activates a nonselective cation channel in *Xenopus laevis* oocytes. Pflügers Archiv. 1998;**436**:329-337. DOI: 10.1007/PL00008085

[99] Schilling WP, Sinkins WG, Estacion M. Maitotoxin activates a nonselective cation channel and a P2Z/$P2X_7$-like cytolytic pore in human skin fibroblasts. American Physiological Society Journal. 1999;**277**:755-765. DOI: 10.1152/ajpcell.1999.277.4.C755

[100] Lundy PM, Nelson P, Mi L, Frew R, Minaker S, Vair C, et al. Pharmacological differentiation of the $P2X_7$ receptor and the maitotoxin-activated cationic channel. European Journal of Pharmacology. 2004;**487**:17-28. DOI: 10.1016/j.ejphar.2004.01.010

[101] Faria RX, Reis RAM, Casabulho CM, Alberto AVP, de Farias FP, Henriques-Pons A, et al. Pharmacological properties of a pore induced by raising intracellular Ca^{2+}. American Physiological Society Journal. 2009;**297**:28-42. DOI: 10.1152/ajpcell.00476.2008

[102] Gaete P, Lillo M, Lopez W, Liu Y, Harris A, Contreras J. A novel voltage clamp/dye uptake assay reveals saturable transport of molecules through CALHM1 and connexin channels. The Journal of General Physiology. 2020;**152**:1-15. DOI: 10.1085/jgp.202012607

[103] Herve CJ, Sarrouilhe D. Connexin-made channels as pharmacological targets. Current Pharmaceutical Design. 2005;**11**:1941-1958. Available from: http://benthamscience.com/article/6595

[104] Ayad WA, Locke D, Koreen IV, Harris AL. Heteromeric, but not homomeric, connexin channels are selectively permeable to inositol phosphates. The Journal of Biological Chemistry. 2006;**281**:16727-16739. DOI: 10.1074/jbc.M600136200

[105] Carnarius C, Kreir M, Krick M, Methfessel C, Moehrle V, Valerius O, et al. Green fluorescent protein changes the conductance of connexin 43 (Cx43) hemichannels reconstituted in planar lipid bilayers. The Journal of Biological Chemistry. 2012;**287**:2877-2886. DOI: 10.1074/jbc.M111.319871

[106] Sáez JC, Vargas AA, Hernández DE, Ortiz FC, Giaume C, Orellana JA. Permeation of molecules through astroglial connexin 43 hemichannels is modulated by cytokines with parameters depending on the permeant species. International Journal of Molecular Sciences. 2020;**21**:1-22. DOI: 10.3390/ijms21113970

[107] Steinberg TH, Newman AS, Swanson JA, Silverstein SC. ATP^{4-} permeabilizes the plasma membrane of mouse macrophages to fluorescent dyes. The Journal of Biological Chemistry.

1987;**262**:8884-8888. DOI: 10.1016/S0021-9258(18)47497-2

[108] Alves LA, Coutinho-Silva R, Persechini PM, Spray DC, Savino W, de Carvalho ACC. Are there functional gap junctions or junctional hemichannels in macrophages? Blood. 1996;**88**:328-334

[109] Alberto AVP, Faria RX, Couto CGC, Ferreira LGB, Souza CAM, Teixeira PCN, et al. Is pannexin the pore associated with the $P2X_7$ receptor? Naunyn-Schmiedeberg's Archives of Pharmacology. 2013;**386**:775-787. DOI: 10.1007/s00210-013-0868-x

[110] Ding S, Sachs F. Inactivation of $P2X_2$ purinoceptors by divalent cations. The Journal of Physiology. 2000;**522**:199-214. DOI: 10.1111/j.1469-7793.2000.t01-1-00199.x

[111] Li M, Toombes GES, Silberberg SD, Swartz KJ. Physical basis of apparent pore dilation of ATP-activated P2X receptor channels. Nature Neuroscience. 2015;**18**:1577-1583. DOI: 10.1038/nn.4120

[112] Wei L, Caseley E, Li D, Jiang L-H. ATP-induced P2X receptor-dependent large pore formation: How much do we know? Frontiers in Pharmacology. 2016;**7**:1-4. DOI: 10.3389/fphar.2016.00005

[113] Stokes L, Layhadi JA, Bibic L, Dhuna K, Fountain SJ. $P2X_4$ receptor function in the nervous system and current breakthroughs in pharmacology. Frontiers in Pharmacology. 2017;**8**:1-15. DOI: 10.3389/fphar.2017.00291

[114] Mafra JCM, Boechat N, Teixeira GP, Faria RX. Synthetic molecules as P2X7 receptor antagonists: A medicinal chemistry update focusing the therapy of inflammatory diseases. European Journal of Pharmacology. 2023;**957**:175999. DOI: 10.1016/j.ejphar.2023.175999

[115] Di Virgilio F, Dal Ben D, Sarti AC, Giuliani AL, Falzoni S. The P2X7 receptor in infection and inflammation. Immunity. 2017;**47**:15-31. DOI: 10.1016/j.immuni.2017.06.020

[116] Pacheco PAF, Ferreira LBG, Mendonça L, Ferreira DNM, Salles JP, Faria RX, et al. P2X7 receptor as a novel drug delivery system to increase the entrance of hydrophilic drugs into cells during photodynamic therapy. Journal of Bioenergetics and Biomembranes. 2016;**48**:397-411. DOI: 10.1007/s10863-016-9668-6

[117] Javed S, Koch-Nolte F, Haag F. ATP-gated P2X7 cation channel synergistically potentiates the cytotoxic effect of doxorubicin in lymphoma model (*In vitro*). Clinical Lymphoma. 2018;**18**:289-290. DOI: 10.1016/j.clml.2018.07.230

[118] Karasawa A, Michalski K, Mikhelzon P, Kawate T. The $P2X_7$ receptor forms a dye-permeable pore independent of its intracellular domain but dependent on membrane lipid composition. eLife. 2017;**6**:1-22. DOI: 10.7554/eLife.31186

[119] Petrou S, Ugur M, Drummond RM, Singer JJ, Walsh J. $P2X_7$ purinoceptor expression in *Xenopus oocytes* is not sufficient to produce a pore-forming P2Z-like phenotype. FEBS Letters. 1997;**411**:339-345. DOI: 10.1016/S0014-5793(97)00700-X

[120] Klapperstück M, Büttner C, Böhm T, Schmalzing G, Markwardt F. Characteristics of $P2X_7$ receptors from human B lymphocytes expressed in *Xenopus oocytes*. Biochimica et Biophysica Acta - Biomembrane. 2000;**1467**:444-456. DOI: 10.1016/S0005-2736(00)00245-5

[121] Persechini PM, Bisaggio RC, Alves-Neto JL, Coutinho-Silva R. Extracellular ATP in the lymphohematopoietic system: P2Z purinoceptors off membrane permeabilization. Brazilian Journal of Medical and Biological Research. 1998;**31**:25-34. DOI: 10.1590/S0100-879X1998000100004

[122] Donnelly Roberts DL, Namovic MT, Faltynek CR, Jarvis MF. Mitogen-activated protein kinase and caspase signaling pathways are required for $P2X_7$ receptor ($P2X_7R$)-induced pore formation in human THP-1 cells. The Journal of Pharmacology and Experimental Therapeutics. 2004;**308**:1053-1061. DOI: 10.1124/jpet.103.059600

[123] Pelegrin P, Surprenant A. Pannexin-1 mediates large pore formation and interleukin-1β release by the ATP-gated $P2X_7$ receptor. The EMBO Journal. 2006;**25**:5071-5082. DOI: 10.1038/sj.emboj.7601378

[124] Locovei S, Scemes E, Qiu F, Spray DC, Dahl G. Pannexin1 is part of the pore forming unit of the $P2X_7$ receptor death complex. FEBS Letters. 2007;**581**:483-488. DOI: 10.1016/j.febslet.2006.12.056

[125] Yan Z, Li S, Liang Z, Tomić M, Stojilkovic SS. The $P2X_7$ receptor channel pore dilates under physiological ion conditions. The Journal of General Physiology. 2008;**132**:563-573. DOI: 10.1085/jgp.200810059

[126] Faria RX, Ferreira LGB, Alves LA. The mystery of $P2X_7$ ionotropic receptor: From a small conductance channel to a large conductance channel. In: Contreras CM, editor. Neuroscience—Dealing with Frontiers. Rijeka: IntechOpen; 2012. pp. 1-18. DOI: 10.5772/29146

[127] MacVicar BA, Thompson RJ. Nonjunction functions of pannexin-1 channels. Trends in Neurosciences. 2010;**33**:93-102. DOI: 10.1016/j.tins.2009.11.007

[128] Sandilos JK, Bayliss DA. Physiological mechanisms for the modulation of pannexin 1 channel activity. The Journal of Physiology. 2012;**590**:6257-6266. DOI: 10.1113/jphysiol.2012.240911

[129] Nielsen BL, Toft-Bertelsen T, Lolansen S, Anderson C, Nielsen M, Thompson R, et al. Pannexin 1 activation and inhibition is permeant-selective. The Journal of Physiology. 2020;**598**:361-379. DOI: 10.1113/JP279263

[130] Loureiro A, Moura-Neto L, Martins C, Silva P, Lopes M, Leitão R, et al. Role of Pannexin-1-$P2X_7R$ signaling on cell death and pro-inflammatory mediator expression induced by *Clostridioides difficile* toxins in enteric glia. Frontiers in Immunology. 2022;**13**:1-13. DOI: 10.3389/fimmu.2022.956340

[131] Ma W, Hui H, Pelegrin P, Surprenant A. Pharmacological characterization of pannexin-1 currents expressed in mammalian cells. The Journal of Pharmacology and Experimental Therapeutics. 2009;**328**(2):409-418. DOI: 10.1124/jpet.108.146365

[132] De Pinto V, Messina A, Lane DJR, Lawen A. Voltage-dependent anion-selective channel (VDAC) in the plasma membrane. FEBS Letters. 2010;**584**:1793-1799. DOI: 10.1016/j.febslet.2010.02.049

[133] Elinder F, Akanda N, Tofighi R, Shimizu S, Tsujimoto Y, Orrenius S, et al. Opening of plasma membrane voltage-dependent anion channels

(VDAC) precedes caspase activation in neuronal apoptosis induced by toxic stimuli. Cell Death and Differentiation. 2005;**12**:1134-1140. DOI: 10.1038/sj.cdd.4401646

[134] Okada SF, O'Neal WK, Huang P, Nicholas RA, Ostrowski LE, Craigen WJ, et al. Voltage-dependent anion channel-1 (VDAC-1) contributes to ATP release and cell volume regulation in murine cells. The Journal of General Physiology. 2004;**124**:513-526. DOI: 10.1085/jgp.200409154

[135] Yehezkel G, Hadad N, Zaid H, Sivan S, Shoshan-Barmatz V. Nucleotide-binding sites in the voltage-dependent anion channel: Characterization and localization. The Journal of Biological Chemistry. 2006;**281**:5938-5946. DOI: 10.1074/jbc.M510104200

[136] Sabirov RZ, Merzlyak PG, Islam MR, Okada T, Okada Y. The properties, functions, and pathophysiology of maxi-anion channels. Pflügers Archiv: European Journal of Physiology. 2016;**468**:405-420. DOI: 10.1007/s00424-015-1774-5

[137] Blatz AL, Magleby KL. Single chloride-selective channels active at resting membrane potentials in cultured rat skeletal muscle. Biophysical Journal. 1983;**47**:119-123. DOI: 10.1016/s0006-3495(85)83884-4

[138] Hyllienmark L, Brismar T. Effect of hypoxia on membrane potential and resting conductance in rat hippocampal neurons. Neuroscience. 1999;**91**:511-517. DOI: 10.1016/S0306-4522(98)00650-2

[139] Gönczi M, Szentandrássy N, Fülöp L, Telek A, Szigeti GP, Magyar J, et al. Hypotonic stress influence the membrane potential and alter the proliferation of keratinocytes in vitro. Experimental Dermatology. 2007;**16**:302-310. DOI: 10.1111/j.1600-0625.2006.00533.x

[140] Toychiev AH, Sabirov RZ, Takahashi N, Ando-Akatsuka Y, Liu H, Shintani T, et al. Activation of maxi-anion channel by protein tyrosine dephosphorylation. American Physiological Society Journal. 2009;**297**:990-1000. DOI: 10.1152/ajpcell.00131.2009

[141] Ferreira LGB, Faria RX. TRPing on the pore phenomenon: What do we know about transient receptor potential ion channel-related pore dilation up to now? Journal of Bioenergetics and Biomembranes. 2016;**48**:1-12. DOI: 10.1007/s10863-015-9634-8

[142] Nagy I, Friston D, Valente JS, Perez JVT, Andreou AP. Pharmacology of the capsaicin receptor, transient receptor potential vanilloid type-1 ion channel BT—Capsaicin as a therapeutic molecule. In: Abdel-Salam OME, editor. Progress in Drug Research. Basel: Springer Basel; 2014. pp. 39-76. DOI: 10.1007/978-3-0348-0828-6_2

[143] Lee YW, Won C-H, Jung K, Nam H-J, Choi G, Park Y-H, et al. Efficacy and safety of PAC-14028 cream—A novel, topical, nonsteroidal, selective TRPV1 antagonist in patients with mild-to-moderate atopic dermatitis: A phase IIb randomized trial. The British Journal of Dermatology. 2019;**180**:1030-1038. DOI: 10.1111/bjd.17455

[144] Park CW, Kim BJ, Lee YW, Won C, Park CO, Chung BY, et al. Asivatrep, a TRPV1 antagonist, for the topical treatment of atopic dermatitis: Phase 3, randomized, vehicle-controlled study (CAPTAIN-AD). The Journal of Allergy and Clinical Immunology. 2022;**149**:1340-1347. DOI: 10.1016/j.jaci.2021.09.024

[145] Veldhuis NA, Lew MJ, Abogadie FC, Poole DP, Jennings EA, Ivanusic JJ, et al. N-glycosylation determines ionic permeability and desensitization of the TRPV1 capsaicin receptor. The Journal of Biological Chemistry. 2012;**287**:21765-21772. DOI: 10.1074/jbc.M112.342022

[146] Tominaga M, Moriyama T. Functional interaction between ATP and TRPV1 receptors BT—Molecular sensors for cardiovascular homeostasis. In: Wang DH, editor. Molecular Sensors for Cardiovascular Homeostasis. Boston, MA: Springer US; 2007. pp. 133-140. DOI: 10.1007/978-0-387-47530-1_7

[147] Myrdal SE, Steyger PS. TRPV1 regulators mediate gentamicin penetration of cultured kidney cells. Hearing Research. 2005;**204**:170-182. DOI: 10.1016/j.heares.2005.02.005

[148] Lee J-H, Park C, Kim S-J, Kim H-J, Oh G-S, Shen A, et al. Different uptake of gentamicin through TRPV1 and TRPV4 channels determines cochlear hair cell vulnerability. Experimental & Molecular Medicine. 2013;**45**:1-13. DOI: 10.1038/emm.2013.25

[149] Biggs JE, Stemkowski PL, Knaus EE, Chowdhury MA, Ballanyi K, Smith PA. Suppression of network activity in dorsal horn by gabapentin permeation of TRPV1 channels: Implications for drug access to cytoplasmic targets. Neuroscience Letters. 2015;**584**:397-402. DOI: 10.1016/j.neulet.2014.07.033

[150] Lim TKY, MacLeod BA, Ries CR, Schwarz SKW. The quaternary lidocaine derivative, QX-314, produces long-lasting local anesthesia in animal models in vivo. Anesthesiology. 2007;**107**:305-311. DOI: 10.1097/01.anes.0000270758.77314.b4

[151] Stueber T, Eberhardt MJ, Hadamitzky C, Jangra A, Schenk S, Dick F, et al. Quaternary lidocaine derivative QX-$_{314}$ activates and permeates human TRPV1 and TRPA1 to produce inhibition of sodium channels and cytotoxicity. Anesthesiology. 2016;**124**:1153-1165. DOI: 10.1097/ALN.0000000000001050

[152] Peters CM, Ririe D, Houle TT, Aschenbrenner CA, Eisenach JC. Nociceptor-selective peripheral nerve block induces delayed mechanical hypersensitivity and neurotoxicity in rats. Anesthesiology. 2014;**120**:976-986. DOI: 10.1097/ALN.0000000000000088

[153] Leffler A, Fischer MJ, Rehner D, Kienel S, Kistner K, Sauer SK, et al. The vanilloid receptor TRPV1 is activated and sensitized by local anesthetics in rodent sensory neurons. The Journal of Clinical Investigation. 2008;**118**:763-776. DOI: 10.1172/JCI32751

[154] Binshtok AM, Gerner P, Oh SB, Puopolo M, Suzuki S, Roberson DP, et al. Coapplication of lidocaine and the permanently charged sodium channel blocker QX-314 produces a long-lasting nociceptive blockade in rodents. Anesthesiology. 2009;**111**:127-137. DOI: 10.1097/ALN.0b013e3181a915e7

[155] Morelli MB, Offidani M, Alesiani F, Discepoli G, Liberati S, Olivieri A, et al. The effects of cannabidiol and its synergism with bortezomib in multiple myeloma cell lines. A role for transient receptor potential vanilloid type-2. International Journal of Cancer. 2014;**134**:2534-2546. DOI: 10.1002/ijc.28591

[156] Nabissi M, Morelli MB, Santoni M, Santoni G. Triggering of the TRPV2 channel by cannabidiol sensitizes glioblastoma cells to cytotoxic chemotherapeutic agents. Carcinogenesis. 2013;**34**:48-57. DOI: 10.1093/carcin/bgs328

[157] Steyger PS, Karasawa T. Intracochlear trafficking of aminoglycosides. Communicative & Integrative Biology. 2008;**1**:140-142. DOI: 10.4161/cib.1.2.6888

[158] Earley S, Heppner TJ, Nelson MT, Brayden JE. TRPV4 forms a novel Ca^{2+} signaling complex with ryanodine receptors and BKCa channels. Circulation Research. 2005;**97**:1270-1279. DOI: 10.1161/01.RES.0000194321.60300.d6

[159] Stepanyan RS, Indzhykulian AA, Vélez-Ortega AC, Boger ET, Steyger PS, Friedman TB, et al. TRPA1-mediated accumulation of aminoglycosides in mouse cochlear outer hair cells. Journal of the Association for Research in Otolaryngology. 2011;**12**:729-740. DOI: 10.1007/s10162-011-0288-x

[160] Estacion M, Schilling WP. Maitotoxin-induced membrane blebbing and cell death in bovine aortic endothelial cells. BMC Physiology. 2001;**1**:1-12. DOI: 10.1186/1472-6793-1-2

[161] Treviño CL, De la Vega-Beltrán JL, Nishigaki T, Felix R, Darszon A. Maitotoxin potently promotes Ca^{2+} influx in mouse spermatogenic cells and sperm, and induces the acrosome reaction. Journal of Cellular Physiology. 2006;**206**:449-456. DOI: 10.1002/jcp.20487

[162] Xi D, Van Dolah FM, Ramsdell JS. Maitotoxin induces a calcium-dependent membrane depolarization in GH4C1 pituitary cells via activation of type L voltage-dependent calcium channels. The Journal of Biological Chemistry. 1992;**267**:25025-25031. DOI: 10.1016/S0021-9258(19)74000-9

[163] Gutierrez D, Díaz de León L, Vaca L. Characterization of the maitotoxin-induced calcium influx pathway from human skin fibroblasts. Cell Calcium. 1997;**22**:31-38. DOI: 10.1016/S0143-4160(97)90087-7

[164] Yue L, Xu H. TRP channels in health and disease at a glance. Journal of Cell Science. 2021;**134**:1-10. DOI: 10.1242/jcs.258372

[165] Flores PL, Rodríguez E, Zapata E, Carbó R, Farías JM, Martínez M. Maitotoxin is a potential selective activator of the endogenous transient receptor potential canonical type 1 channel in *Xenopus laevis* oocytes. Marine Drugs. 2017;**15**:1-11

[166] Brereton HM, Chen J, Rychkov G, Harland ML, Barritt GJ. Maitotoxin activates an endogenous nonselective cation channel and is an effective initiator of the activation of the heterologously expressed hTRPC-1 (transient receptor potential) nonselective cation channel in H4-IIE liver cells. Biochimica et Biophysica Acta - Biomembranes. 2001;**1540**:107-126

[167] Rychkov G, Barritt GJ. TRPC1 Ca^{2+}-permeable channels in animal cells BT—Transient receptor potential (TRP) channels. In: Flockerzi V, Nilius B, editors. Handbook of Experimental Pharmacology. Vol. 179. Berlin, Heidelberg: Springer; 2007. pp. 23-52. DOI: 10.1007/978-3-540-34891-7_2

[168] Liu C, Hermann T. Characterization of ionomycin as a calcium ionophore. The Journal of Biological Chemistry. 1978 Oct;**1**(253):5892-5894. DOI: 10.1016/S0021-9258(17)34550-7

[169] Pinton P, Giorgi C, Siviero R, Zecchini E, Rizzuto R. Calcium and apoptosis: ER-mitochondria Ca^{2+} transfer in the control of apoptosis. Oncogene. 2008;**27**:6407-6418. DOI: 10.1038/onc.2008.308

[170] Lollike K, Borregaard N, Lindau M. The exocytotic fusion pore of small granules has a conductance similar to an ion channel. The Journal of Cell Biology. 1995;**129**:99-104. DOI: 10.1083/jcb.129.1.99

[171] Soares-Bezerra RJ, Pinho RT, Bisaggio R d C, Benévolo-de-Andrade TC, Alves LA. The search for new agonists to $P2X_7R$ for clinical use: Tuberculosis as a possible target. Cellular Physiology and Biochemistry. 2015;**37**:409-418. DOI: 10.1159/000430364

Chapter 7

The 'Patch-Clamp' Technique and Its Application in Studies on Voltage-Gated Potassium Channels Kv1.3 in Normal and Cancer Cells

Andrzej Teisseyre, Kamila Środa-Pomianek and Anna Palko-Labuz

Abstract

The introduction of this chapter presents a historical outline of experimental methods applied in electrophysiology until development of the 'patch-clamp' technique. The first section briefly describes the 'patch-clamp' technique and its configurations, with areas of practical applications of the configurations. The second section of the chapter presents an application of the 'whole-cell' configuration in studying activity of voltage-gated potassium channels Kv1.3. It is pointed out that an application of this configuration enabled the discovery of these channels in human T lymphocytes in 1984. Studies performed later applying the 'whole-cell' configuration have shown that these channels are also expressed in many other cell types, both normal and cancer, both in the plasma membrane and in the inner mitochondrial membrane. It was also shown that the channels may be blocked by many chemically unrelated compounds. Finally, results obtained upon a combination of the 'whole-cell' recording with non-electrophysiological techniques provided evidence that some inhibitors of these channels may find a clinical application in therapy of many diseases, including T-cell mediated autoimmune diseases, chronic inflammatory diseases, severe cases of COVID-19 and some cancer disorders.

Keywords: patch-clamp, Kv1.3 channel, autoimmune diseases, chronic inflammatory diseases, COVID-19, cancer disorders

1. Introduction

History of electrophysiology began at the beginning of the XX-th century. In 1902, Julius Bernstein formulated the 'potassium electrode hypothesis'. According to this hypothesis, the resting cell membrane is selectively permeable for potassium ions [1–3]. Thus, the resting membrane potential can be calculated applying the Nernst's equilibrium potential for potassium ions. Taking into account that the intracellular potassium concentration is much higher than the extracellular one, the resting membrane potential is negative and equal to ca. -90 mV. Bernstein proposed that upon an excitation, the membrane is undergoing a transient 'breakdown' becoming equally permeant for all ions. Under these circumstances, the membrane potential is changing from the resting negative value to ca. 0 mV – a membrane is being 'depolarised'. Three years later, Herrman proposed the 'cable model' of an axon having a cylindrical conducting core surrounded by a membrane with a high electrical resistance and capacitance, similarly to a submarine cable [4, 5]. A depolarisation of one point of the cable is propagated as a longitudinal electric current flowing from region being depolarised to regions not yet depolarised [4, 5].

The main problem for electrophysiologists at that time was lack of an experimental technique that could be applied to verify theoretical hypotheses. The first electronic device, a Wheatstone bridge, was applied in electrophysiology only in 1920s. In 1923, Cole and Curtis applied a Wheatstone bridge to measure cell membrane resistance and capacitance. Their results showed that a cell membrane has a high resistance (in the gigaohm range) and capacitance, whereas the cytoplasm has conductance of 30–60% of the conductance of the bathing saline [1]. Results obtained by Curtis and Cole confirmed that the Herrman's 'cable model' was correct.

Electrophysiological studies on electrical properties of cell membranes were pushed forward after a rediscovery of a squid giant axon by Young in 1936 [6]. It has become a valuable model system for further electrophysiological studies. Results published by Hodgkin in 1937 showed that a depolarisation wave during propagation of action potentials in axons was a longitudinal electric current. These results were in accordance to the Herrman's 'cable model' [7]. In 1939, changes of membrane conductance upon firing action potentials in a squid giant axon were measured by Cole and Curtis [8]. The same year, Hodgkin and Huxley measured full action potentials in squid giant axons [9]. Results obtained by Hodgkin and Huxley were confirmed in studies performed independently by Curtis and Cole [10–11]. These results provided evidence that resting membrane potential in an axon was close to the Nernst's equilibrium potential for potassium ions. These results were in accordance to the 'potassium electrode hypothesis'. However, upon an excitation, no 'breakdown' of the cell membrane was observed [8–11]. Instead, the cell membrane was transiently being depolarised to ca. +40 mV [9–11]. Thus, the membrane potential changed not only its value but also its sign from negative to positive. Such a change of the sign was an 'overshoot' [9–11]. The mystery of this 'overshoot' remained unexplained for several years.

Studies on the nature of action potentials in squid giant axons were continued in the late 1940s by Hodgkin, Huxley and Katz. Results of their experiments provided evidence that the 'overshoot' was due to a transiently increased membrane permeability for sodium ions [12]. Because the extracellular sodium concentration is much higher than the intracellular one, the Nernst's equilibrium potential for sodium ions is positive (ca. +50 mV). Thus, increase in the membrane permeability for sodium ions caused a massive influx of these ions changing the membrane potential to a value close to the Nernst equilibrium potential for sodium ions. This 'sodium hypothesis' provided a correct explanation for the depolarisation phase of action potentials in excitable cells [12].

The next important step in electrophysiology was a development of a new experimental technique. The 'voltage-clamp' was developed independently by Marmont [13], Cole [14] and then by Hodgkin, Huxley and Katz (1949–1952) [15–18]. During 'voltage-clamp' experiments, ion currents flowing through the selected cell

DOI: http://dx.doi.org/10.5772/intechopen.1006991

membrane are measured under a controlled ('clamped') voltage. The voltage can be kept constant or changed during an experiment, according to the experimental protocol applied by a researcher. In other words, the 'voltage-clamp' enables measurements of ion currents flowing in response to a given voltage stimulus.

The idea of the 'voltage-clamp' measurement is briefly presented on **Figure 1A**. The membrane potential E is measured by the voltage microelectrode – E`. The electrode is connected to the high-impedance follower circuit and to the input of the amplifier working in a negative feedback loop. The command voltage (V_{comm}) is applied to the amplifier, which compares it to the membrane potential E. If the E is different from the V_{comm}, the amplifier compensates for the difference immediately by applying the error signal. The error signal is the current I, which is applied to the examined cell by the current microelectrode I`. The current I changes the membrane potential to a value of the V_{comm}. Thus, the current recorded in a 'voltage-clamp' measurement is the error signal. An electrical circuit is closed by an additional grounded current microelectrode I″.

An equivalent electric circuit of the cell membrane is shown in **Figure 1B**. The circuit contains voltage-dependent sodium and potassium conductance, g_{Na} and g_K, respectively; voltage-independent leak conductance g_L and membrane capacitance, C_M.

Currents recorded upon an application of a given voltage stimulus were a sum of sodium and potassium currents, leak current and capacitance current. Both the leak and the capacitance components were irrelevant for electrophysiological studies and were subtracted from the final record. Ion currents recorded from the giant axon of *Loligo* upon the membrane depolarisation contained a transient inward sodium current followed by a sustained outward potassium current. After having applied an

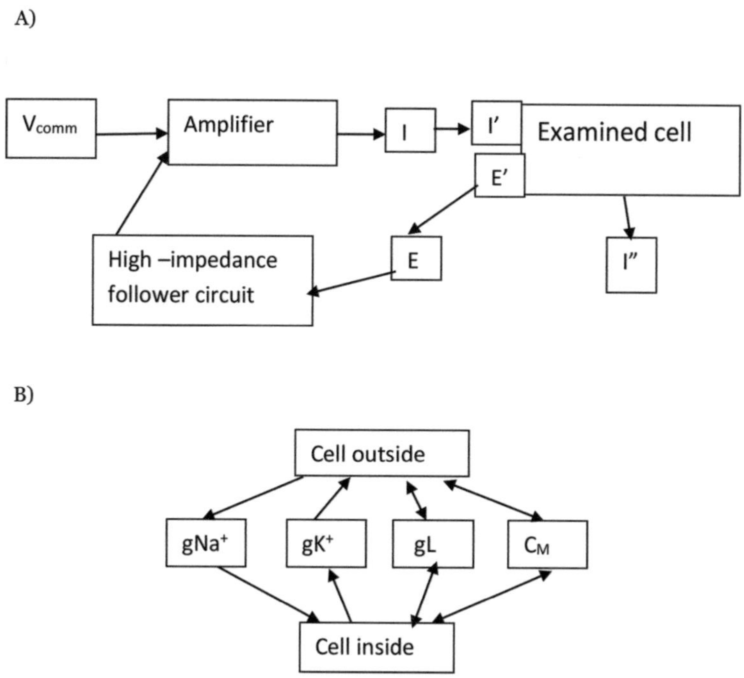

Figure 1.
(A) Scheme of the electric circuit applied in the 'voltage-clamp' technique, and (B) Equivalent circuit of the cell membrane. A detailed description in the text.

extracellular sodium-free solution, only a sustained outward potassium current was recorded. The sodium current was obtained as the difference current [15].

Hodgkin and Huxley also built a theoretical model, which correctly describes changes in ion permeability when firing action potentials in electrically excitable cells. This 'Hodgkin-Huxley Model' is still used for computer simulations of action potentials, in both research studies and teaching activities.

According to the model, the membrane permeability for sodium and potassium ions is represented by the sodium and potassium conductance, respectively [19]. Both conductances depend on the membrane potential and time. When the action potential is being generated, the sodium conductance is rapidly rising to the maximum, due to activation of voltage-gated sodium channels. After a short time, it is decaying to zero, due to the channel inactivation. Inactivated channels are unable to conduct ions even while being open. The channels have to recover from inactivation in order to be able to work again. The potassium conductance is rising slower, due to a delayed activation of 'delayed rectifier' voltage-gated potassium channels. According to the Hodgkin-Huxley Model, the 'delayed rectifier' voltage-gated potassium channels do not inactivate. Therefore, the potassium channels remain active for a longer time and the potassium conductance remains high until the channels finally close at the end of the hyperpolarisation phase of action potential. A rapid activation of voltage-gated sodium channels (working in a positive feedback loop) is responsible for the depolarisation phase of action potential. A combination of inactivation of voltage-gated sodium channels and delayed activation of voltage-gated potassium channels (working in a negative feedback loop) is responsible for the repolarisation phase, when the membrane potential is going back to the resting value. Long-lasting activity of 'delayed rectifier' voltage-gated potassium channels is responsible for the hyperpolarisation phase, which follows the repolarisation. During hyperpolarisation, the membrane potential is below the resting value being almost equal to the Nernst's equilibrium potential for potassium ions. The membrane hyperpolarisation is necessary to make voltage-gated sodium channels recover from inactivation. Recovery from inactivation requires some time. During that time known as time of the absolute refraction, no action potential can be generated even when a strong stimulatory pulse is applied. The period of absolute refraction is followed by a period of relative refraction when a strong stimulatory pulse is able to generate next action potential, whereas a standard pulse still cannot.

The activation and inactivation of voltage-gated sodium channels, as well as the activation and closure of voltage-gated potassium channels, were described applying a first-order chemical reaction formalism. An ion current flowing through a membrane is a sum of currents flowing through single ion channels.

The Hodgkin-Huxley Model correctly described opening of single ion channels, which is a stochastic process. The probability of channel opening depends on membrane potential and time. The channels are open by so-called 'gating particles'. There are three 'gating particles' in case of voltage-gated sodium channels and four gating particles in case of 'delayed rectifier' voltage-gated potassium channels. The channels are open only if all the gating particles are in the 'open' position. If the probability of opening of one gating particle of a voltage-gated sodium channel is equal to m, the probability of opening of the channel is equal to m^3. If the probability of opening of one gating particle of a voltage-gated potassium channel is equal to n, the probability of opening of the channel is equal to n^4. The inactivation of a voltage-gated sodium channel is controlled by a different 'inactivating particle' working independently to the 'gating particles'. If the probability that a sodium channel is not inactivated is equal to h, the probability that the channel is open and not inactivated is equal to m^3h. The values of m, n and h depend on membrane potential and time.

Hodgkin and Huxley did not have any opportunity to measure ion currents flowing through single ion channels, since such measurements were impossible by applying the 'voltage clamp' technique. Single-channel recording has become possible only after having developed the 'patch-clamp' technique, more than 20 years later.

2. The 'patch-clamp' technique: historical outline

The 'patch-clamp' technique was firstly applied by Sakmann and Neher, who wanted to record sodium currents in denervated frog muscle fibres [20]. Their idea was to isolate a 'patch' of the cell membrane. The area of the 'patch' had to be small enough to contain single ion channels. The 'patch' was isolated by applying a glass pipette (with the recording electrode inside) having a narrow tip of a diameter 3–5 μm. The tip was obtained by applying a pipette puller. The tip was pressed against the surface of the examined cell, making a high-resistance (ca. 50 MΩ) electrical contact ('seal') with the 'patch' (the rest of the examined cell remains untouched). Under these experimental conditions, currents flowing through single ion channels were recorded at a given membrane potential. However, these currents were contaminated by a large background noise caused by the 'seal' [20, 21].

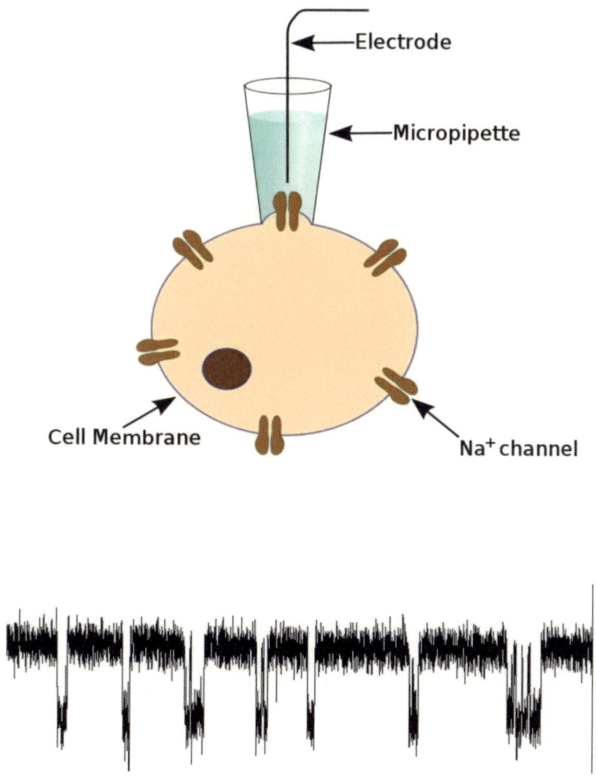

Figure 2.
Above: a scheme of the 'patch-clamp' technique, based on a formation of a 'gigaohm seal', between the tip of the recording pipette and the patch of membrane of the examined cell. Below: an example of an ion current flowing through a single voltage-gated sodium channel (downward deflections).

The 'patch-clamp' technique was significantly improved when the gigaohm-seal ('giga-seal') was discovered in 1980 by Sigworth and Neher (**Figure 2**) [22]. The researchers showed that upon a transient application of a gentle suction to the pipette, a tight contact between the tip and the 'patch' was obtained. It was revealed by a rapid increase of the membrane resistance from megaohm to gigaohm range ('giga-seal'). When a 'giga-seal' was obtained, the 'patch' formed an Ω-shaped protrusion inside the tip. The distance between the rim of the tip and the contacting 'patch' membrane was ca. 1 Å [22]. Such a short distance occurs upon a transfer of lipid monolayers onto glass substrates [23].

An application of 'giga-seals' reduced the background noise in the 'patch' by an order of magnitude [24]. As a result, both quality of records and time resolution of the method were significantly improved [24]. Due to a high resistance of 'giga-seals', a single microelectrode is applied in the 'patch-clamp', to both 'clamp' the membrane potential and record ion currents [22, 24]. Moreover, 'giga-seals' share a high mechanical stability. Therefore, the 'patch' can be excised from the examined cell. Alternatively, it is possible to disrupt the 'patch' and obtain a direct contact with the whole examined cell (see below). Similarl to the 'voltage-clamp' technique, the examined cell must be placed in a bathing solution, with a grounded reference microelectrode, to close the electrical circuit.

3. The 'patch-clamp' recording configurations

3.1 The 'cell-attached'

The improved 'patch-clamp' technique and its basic recording configurations were described in detail in a classical work by Hamill and co-workers [24]. Single-channel recording was described in detail later by Sakmann and Neher [25]. A schematic diagram with the configurations is depicted in **Figure 3**.

As it is shown, the basic 'cell attached' configuration is established after having obtained the 'giga-seal'. Under 'cell attached' conditions, single-channel currents from a membrane patch on an intact cell are recorded.

The 'cell attached' is the least invasive configuration of the patch-clamp technique. The voltage on the examined 'patch' is equal to the cell resting membrane potential minus the pipette potential. The voltage is 'clamped' provided that the resting cell membrane potential remains stable, while ion currents are flowing through the 'patch'. This is the case when the ion conductance of the 'patch' is much lower than the conductance of the rest of examined cell.

The ability to record single-channel currents from the 'patch' is limited by the background noise. This noise is produced by the 'giga-seal', by leak currents flowing through the 'patch', by the recording pipette and by the electronic circuit [24]. The background noise limits suitability of the 'cell-attached' configuration to ion channels having the single-channel conductance more than a few pS [24].

3.2 The 'inside-out'

The 'patch' can be excised from the rest of the examined cell by withdrawal the recording pipette (**Figure 3**). After the excision, a tight vesicle is formed on the pipette tip [24]. The vesicle contains two membranes: the inner and the outer one [24]. In order to record ion currents from the 'patch', the outer membrane has to be

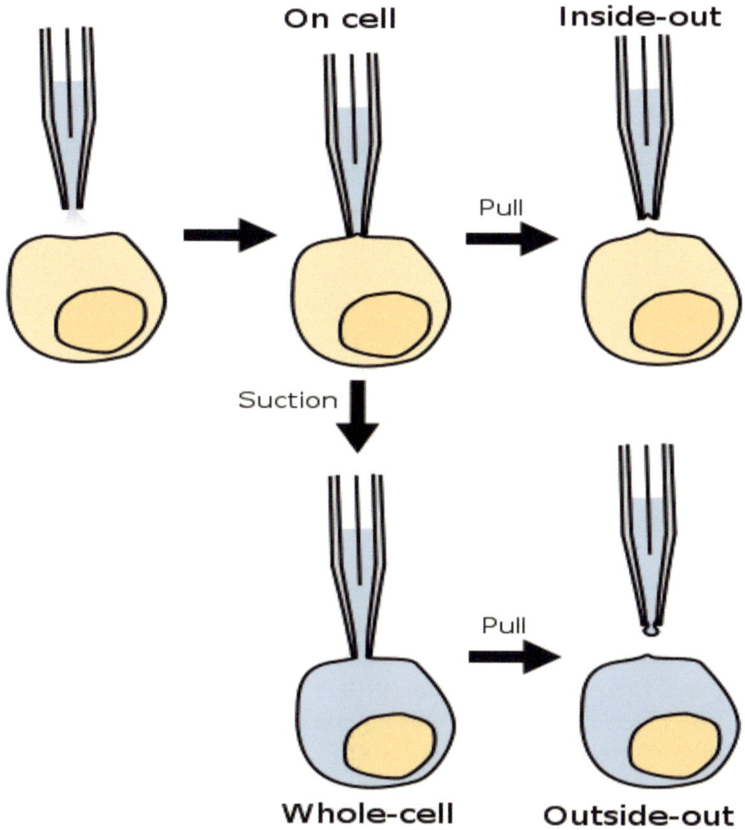

Figure 3.
Recording configurations of the 'patch-clamp' technique based on a formation of a 'giga-seal'.

disrupted. This can be done by withdrawal the recording pipette from the bathing solution for a short time, or by applying a calcium-free bathing solution [24].

The intracellular side of the 'patch' is directly exposed to the bathing solution. This configuration is useful for studying ionic channels activated by intracellular factors, such as intracellular calcium. The 'inside-out' configuration is widely applied in studying properties of calcium-activated potassium channels – K(Ca). These channels are open not by changes of the membrane potential, but by intracellular calcium ions. The probability of opening of K(Ca) channels strongly depends on the intracellular calcium concentration. Usually, it is rising from 0 to 100 percent when the intracellular calcium concentration is rising from 100 nM to 1 μM.

In the 'inside-out' configuration, the voltage on the examined 'patch' is equal to the pipette potential with an opposite sign. The 'patch' membrane potential is more stable than in the 'cell-attached' configuration. On the other hand, the 'inside-out' configuration is much more invasive. This may cause a wash-out of important intracellular factors controlling the activity of examined channels. Therefore, single-channel characteristics in the 'inside-out' recording may be different than those obtained on intact cells.

3.3 The 'whole-cell'

As it was mentioned above, the 'patch' can be disrupted. This can be done by applying a pipette solution containing 150 mM KCl and low calcium (less than 1 μM), a gentle suction or by a short voltage pulse (ca. 200 mV) (**Figure 3**). After the disruption, a direct contact between the pipette tip and the intracellular milieu is obtained. Importantly, the disruption does not destroy the seal between the pipette rim and the cell membrane [24].

In the 'whole-cell' configuration, the membrane potential of the examined cell is equal to the pipette potential. Area of application of this configuration is limited to small cells (less than 30 μm in diameter) making ion currents not larger than a few nA [24]. The 'whole-cell' recording is irreplaceable for studying ion channels in small cells, such as lymphocytes or lymphoid cell lines (see below). Another important advantage of the 'whole-cell' configuration is ability to achieve the single-channel resolution [24, 26]. By applying the fluctuation analysis, it is possible to calculate the single-channel conductance. Such an approach is irreplaceable for studying ion channels with a conductance less than a few pS. The main disadvantage of the 'whole-cell' recording is its invasiveness, which may cause a wash-out of important intracellular factors controlling the activity of examined channels.

3.4 The 'outside-out'

The 'outside-out' (known also as the 'excised patch') can be obtained from the 'whole-cell' configuration by withdrawal the recording pipette from the examined cell (**Figure 3**). In such a case, a small 'patch' is formed on the pipette tip [24]. The 'patch' is exposed to the bathing solution with its extracellular side. The membrane potential of the 'patch' cell is equal to the pipette potential.

The 'outside-out' recording often follows the 'whole-cell' recording [24]. It is useful for studying channels activated by extracellular ligands, such as neurotransmitters.

Similar, to the 'whole-cell', the main disadvantage of this configuration is its invasiveness, which may cause a wash-out of important intracellular factors controlling the activity of examined channels.

4. Voltage-gated potassium channels Kv1.3 in normal and cancer cells: studies with an application of the 'patch-clamp' technique in a combination with non-electrophysiological approaches

Voltage-gated potassium channels (Kv channels) are integral membrane proteins that share selectivity for potassium ions and are activated upon a change of the membrane potential to more positive values (membrane depolarisation). Opening of Kv channels provides transportation of potassium ions across the cell membrane down their electrochemical gradient. Movement of potassium ions through the channels makes 'potassium current', which can directly be recorded applying the 'patch-clamp' technique. The direction of these currents depends on the membrane potential. If the potential is higher than the 'reversal potential', the current is outward (positive). If the potential is lower than the 'reversal potential', the current is inward (negative). If the potential is equal to the 'reversal potential', the current is equal to zero. The 'reversal potential' for Kv1.3 channels in the plasma membrane is about −80 mV under physiological conditions. The membrane depolarisation always leads to more positive potentials. Therefore, currents flowing through Kv1.3 channels in the plasma membrane are outward (positive). Kv1.3 channels were firstly described as the 'n' ('normal') channels in 1984 in the plasma membrane in human T lymphocytes [27].

The channels were characterised in detail one year later by Cahalan and co-workers [28]. The authors applied the 'whole-cell' recording to record the channel activity on both the whole-cell and the single-channels level [28]. The channels are 'delayed rectifier' Kv channels, which activate upon the membrane depolarisation and then undergo an inactivation. An 'inactivated' channel is unable to conduct ions even while being open. The channel has to 'recover' from inactivation to become able to work again. It was shown that Kv1.3 channels undergo a complex inactivation, with the 'fast' and the 'slow' component [28–30]. The 'slow' component, originally named as a 'cumulative inactivation', is a characteristic feature of Kv1.3 channels [28]. Recovery from inactivation is also complex, having the 'fast' and the 'slow' component [28]. It was shown that a small number of Kv1.3 channels in the plasma membrane are active under a resting membrane potential. Activation of the channels provides stabilisation of the resting membrane potential [28–30]. Kv1.3 channels work in a negative feedback loop, since opening of the channels counteracts the membrane depolarisation, which leads to the opening. Studies applying non-electrophysiological techniques showed that Kv1.3 channel is one of mammalian *Shaker* Kv channels. Kv1.3 channels are encoded by KCNA3 gene [29, 30]. Kv1.3 channels are homotetramers containing four α subunits and four intracellular regulatory β subunits [29]. Kv1.3 channels are expressed abundantly in many different cell types [29, 30]. The channels are also expressed in the inner mitochondrial membrane (mito Kv1.3 channels) [30–32]. Single-channel mito Kv1.3 channel currents were recorded applying the 'inside-out' configuration on the channels expressed in so-called mitoplasts [31]. Mitoplasts were obtained by applying an osmotic shock to mitochondria isolated from examined cells [31]. During an osmotic shock, mitochondria are swelling. The swelling makes the outer mitochondrial membrane collapse, thereby exposing the stretched inner mitochondrial membrane to the 'patch-clamp' micropipette [31]. Mito Kv1.3 channels are slightly voltage-dependent. The channels are active at the resting potential of the inner mitochondrial potential (IMM), which is equal to ca. -180 mV. This potential is lower than the 'reversal potential' for mitoKv1.3 channels (ca. 0 mV). Therefore, opening of mito Kv1.3 channels induces an influx of potassium ions, thereby depolarising the inner mitochondrial membrane [30–32].

Activity of Kv1.3 channels is required in a proliferation of Kv1.3 channel-expressing cells [30, 33–38]. It was shown that a blockage of Kv1.3 channels in the plasma membrane inhibits Kv1.3 channel-expressing cell proliferation at the checkpoint between the G1 and the S phase of the cell cycle [30, 33–38]. An involvement of Kv1.3 channels in a regulation of the cell proliferation can be described applying two distinct models: the 'membrane potential model' and the 'voltage sensor model' [30, 38].

Kv1.3 channels also participate in induction of apoptosis of normal and cancer cells that express this channel type [30–32, 34, 36–38].

Apoptosis is a process of a programmed cell death, which occurs in all normal cells. Apoptosis occurs *via* two major pathways: the death receptor and the mitochondrial pathway [30–32, 34, 36–38].

It was shown that Kv1.3 channels are required for apoptosis of Kv1.3 channel-expressing cells. The plasma membrane channels are upregulated upon activation of membrane death receptors by Fas ligands [30, 38]. Activation of the channels provides a sustained efflux of potassium ions and cell shrinkage, which is indispensable for induction of apoptosis [30]. On the other hand, it was shown that an inhibition of mito Kv1.3 channels by the pro-apoptotic protein Bax is required to induce the mitochondrial pathway of apoptosis of Kv1.3 channel-expressing cells [30, 34, 36–38].

The mechanism of involvement of mitoKv1.3 channels in the cell apoptosis was described in detail elsewhere [30, 32, 38].

It was shown that some membrane-permeant small-molecule inhibitors of mitoKv1.3 channels may mimic the pro-apoptotic effect of the Bax protein, thereby leading to the cell apoptosis. These compounds may selectively eliminate Kv1.3 channel-expressing cancer cells while sparing the normal ones. The mechanism is described in detail elsewhere [30, 34, 36–38]. The most promising candidates for a putative clinical application in cancer therapy are 'mitochondriotropic' compounds, which combine a high lipophilicity with a positive charge. These compounds preferentially inhibit mitoKv1.3 channels, without significantly affecting the channels in the plasma membrane [30, 34, 36–38].

It is known that the expression of Kv1.3 may significantly be changed (downregulated or upregulated) both under physiological and pathological conditions (**Table 1**) [30, 33, 34, 36–44].

It is known that an inhibition of Kv1.3 channels by may putatively be beneficial in therapy of various diseases (**Figure 4**) [30, 34, 36–38, 41–44].

Kv1.3 channels are inhibited not only by 'classical' blockers of potassium channels, such as tethraethylammonium (TEA) or 4-ammino-pyridine (4-AP), but also by many other compounds, both inorganic and organic [29–38] (see below).

4.1 Patch-clamp studies on the inhibition of Kv1.3 channels in normal and cancer cells: examples of results obtained on model systems

In order to study inhibitory effects of various drugs on Kv1.3 channels, it is useful to apply a model system, in which the channels are expressed endogenously and predominantly. In case of normal cells, such a system are human T lymphocytes isolated from peripheral blood of healthy donors. In case of cancer cells, a useful model system can be lymphoblastic human T cell line Jurkat [38].

Cell type	Changes in expression and its putative clinical significance	References
Activated naïve and central memory human T lymphocytes (TCM)	Upregulation (ca. 300 channels per cell)	[30, 33, 38]
Activated effector memory human T lymphocytes (TEM)	Strong upregulation (ca. 1500 channels per cell)	[30, 33, 38]
CD8+ T lymphocytes in patients with ovarian cancer	Upregulation. Putatively valuable diagnostic and prognostic biomarker of a malignant ovarian cancer	[39]
Microglia, 'classically' activated by lipopolysaccharide (LPS)	Upregulation	[40]
Mature neoplastic B cells in chronic lymphocytic leukaemia (B-CLL)	Upregulation	[30, 41–44]
Human leukaemic Jurkat T cells	Upregulation	[30, 38]
Colon cancer, smooth muscle cancer, skeletal muscle cancer, lymph node cancer	Upregulation	[30, 34, 38, 41–44]
Breast cancer, pancreas adenocarcinoma, colorectal cancer	Downregulation due to methylation of the promoter region of Kv1.3-encoding gene. Putative diagnostic and prognostic marker	[30, 34, 38, 41–44]

Table 1.
Examples of changes of Kv1.3 channel expression under physiological and pathological conditions.

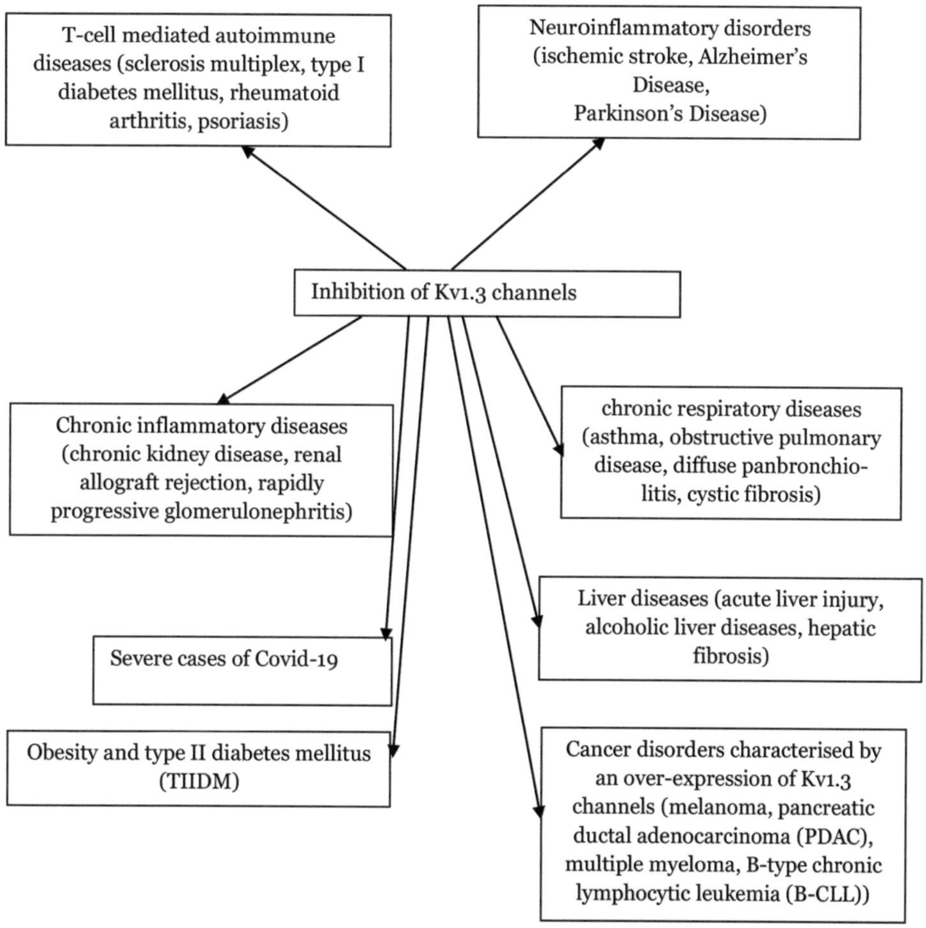

Figure 4.
An overview of diseases, which may putatively be treated by inhibitors of Kv1.3 channels.

Because of a size of both T lymphocytes and Jurkat T cells, it is recommended to apply the whole-cell recording. Two types of experimental protocols can be applied in patch-clamp studies on Kv1.3 channels: sequences of depolarising voltage steps and 'voltage ramps'.

Figure 5 depicts Kv1.3 whole-cell currents recorded in a human T lymphocyte applying a sequence of depolarising voltage steps under control conditions, upon an application of genistein at 40 μM concentration and after wash-out of the drug [45].

The experimental protocol contained 7 depolarising voltage steps from the holding potential of −90 mV, to the potentials from −60 mV to +60 mV, with an increment of 20 mV (**Figure 5**, upper part) [45]. Duration time of each step was 50 ms, the time interval between steps – 30 s [45]. Relatively long time interval between the steps was necessary to provide recovery of the channels from inactivation. Leak current was subtracted online, applying the P/4 procedure [45]. The idea of this procedure is to apply, before each step, four identical pre-pulses of a value equal to ¼ of the applied step and an opposite sign. The currents recorded upon an application of these pre-pulses are then summarised and subtracted from the current recorded upon an

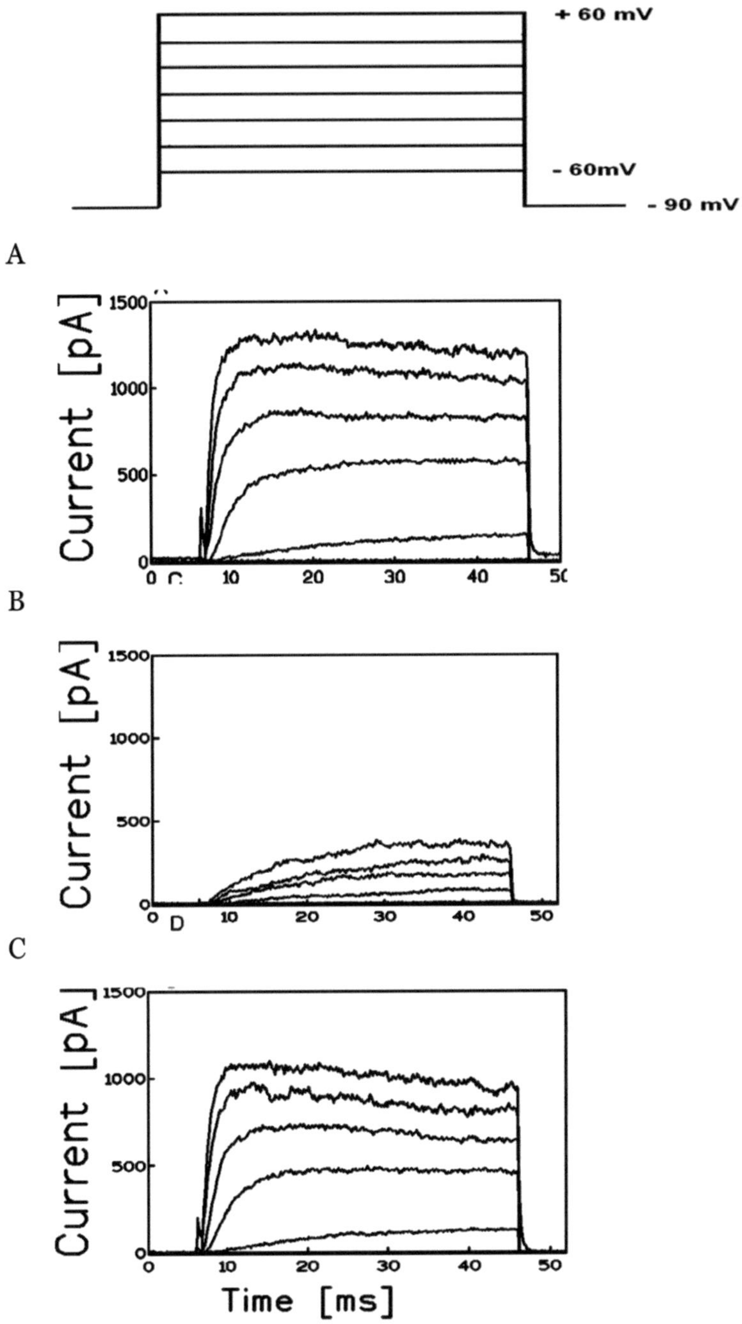

Figure 5.
Top: scheme of a sequence of depolarising voltage steps applied in the experiment. Below: examples of the whole-cell Kv1.3 currents recorded in a normal T lymphocyte under control conditions: (A), upon application, (B) after wash-out; and (C) of genistein at 40 μM concentration.

application of the step. This method enables an online subtraction of the leak current, because this current rises linearly with the voltage, according to the Ohm's law. On the other hand, the Kv1.3 current remains unaffected, since this current is not 'ohmic' (the current-to-voltage relationship is not linear).

Apparently, an application of genistein significantly reduced the currents and slowed down their activation rate. The inhibitory effect was reversible [45].

An alternative approach is an application of the 'voltage ramps', where the cell membrane was gradually depolarised from the holding potential of −90 mV to the potentials from −100 mV (start of the 'ramp') to +40 mV (end of the 'ramp') [46]. Time duration of each 'voltage ramp' was 340 ms, time interval between the 'ramps' - 30 s. Relatively long time interval between the 'ramps' was necessary to provide recovery of the channels from inactivation. **Figure 6** depicts Kv1.3 whole-cell currents recorded in a cancer Jurkat T cell applying 'voltage ramps' under control conditions, upon an application of isobavachalcone (IBC) at 3 μM concentration and after wash-out of the drug [46].

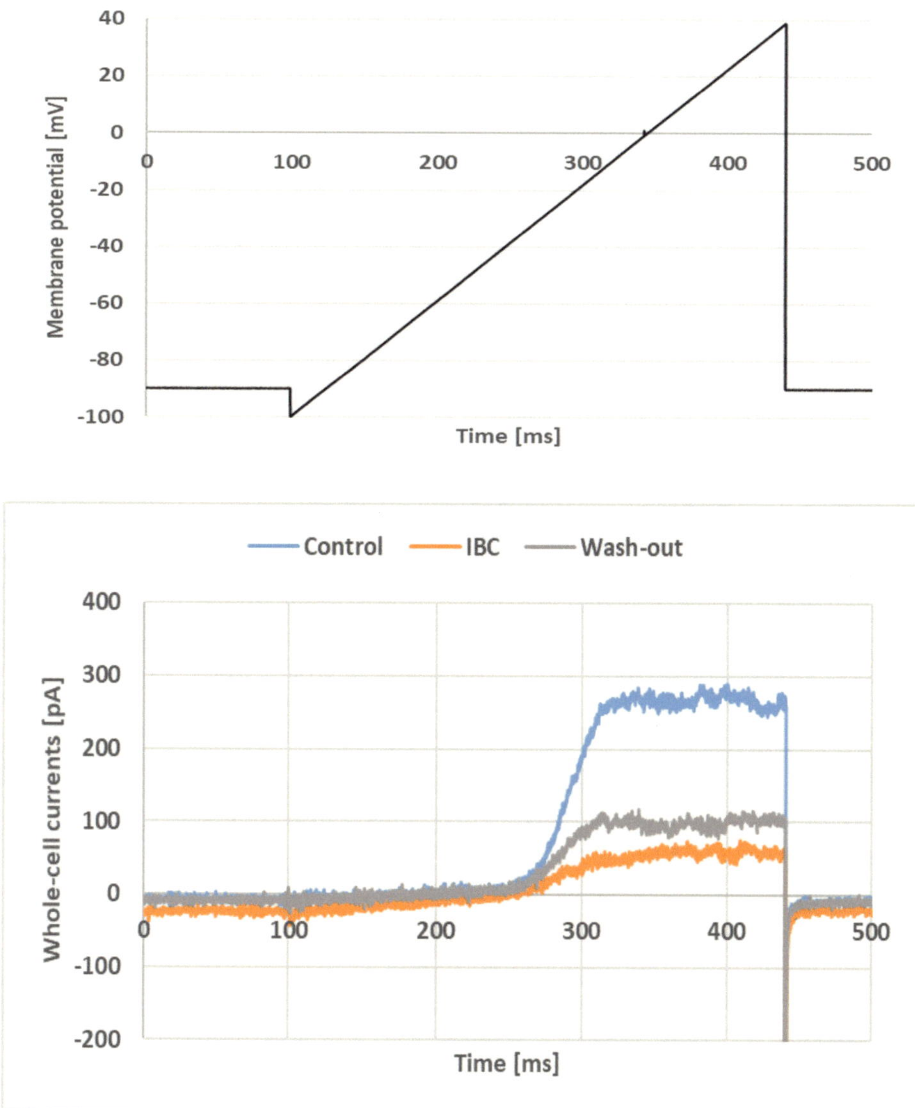

Figure 6.
Whole-cell currents recorded in a Jurkat T cell (lower panel) applying the 'voltage ramp' (upper panel), under control conditions (blue trace), upon an application of 3 μM isobavachalcone (IBC - orange trace) and during wash-out of the drug (grey trace).

The recorded 'ramp current' contained both the Kv1.3 currents and the unspecific leak current (subtracted during a subsequent offline analysis) (**Figure 6**) [46].

Apparently, an application of isobavachalcone significantly reduced the Kv1.3 current [46]. Of note, the Kv1.3 channel component did not recover after wash-out of the drug. This indicates that in this case, the inhibitory effect on Kv1.3 channels was partially irreversible [46].

5. Inhibitors of Kv1.3 channels as putative therapeutic agents

Several groups of Kv1.3 channel inhibitors may putatively be applied in a medicinal practice (**Table 2**). Among them are both peptide inhibitors and small-molecule organic compounds, such as calcium channel blockers, Nonsteroidal Anti-Inflammatory Drugs (NSAIDs), macrolide antibiotics, psoralens and their 'mitochondriotropic' derivatives, naturally occurring polycyclic compounds (**Table 2**). Naturally occurring compounds, which share the ability to inhibit Kv1.3

Class of inhibitors	Names of compounds	Concentration required for channel block	Putative medicinal applications	References
Peptide inhibitors	ShK, BgK, MgTX, BmPO2, BmKTX, OSK1, KTX, ChTX, MTX, NTX, Pi1, Vm24, BF9	From picomolar to nanomolar	T-cell mediated autoimmune diseases, neuroinflam-matory diseases, liver diseases	[34, 35, 38, 47]
Calcium channel blockers	Verapamil, Nifedipine, Diltiazem, Benidipine	Up to 100 μM	Chronic inflammatory diseases	[29, 38, 48, 49]
Nonsteroidal Anti-Inflammatory Drugs, macrolide antibiotics	Diclofenac sodium, salicylate, indomethacin, clarithromycin, chloroquine	Up to 100 μM	Chronic inflammatory diseases, severe cases of COVID-19	[38, 48–50]
Psoralens and clofazimine	Psora-4, PAP-1, clofazimine	Nanomolar	Cancer disorders characterised by an over-expression of Kv1.3 channels	[30, 34, 36, 38]
'Mitochon-Driotropic' derivatives of PAP-1	PAPTP, PCARBTP, PCTP	Low nanomolar, low micromolar	As mentioned above	[30, 36–38, 51–54]
Naturally occurring polycyclic compounds	Flavonoids, polyphenols, chalcones, statins	From low micromolar up to 100 μM	As mentioned above	[30, 38, 55]
Thiophene-based compounds	Series of compounds obtained by structural optimisation of benzamide-based hit compounds	Nanomolar	As mentioned above	[56]

Table 2.
Overview of inhibitors of Kv1.3 channels that may putatively be applied in a medicinal practice.

channels, are flavonoids 8-prenyl- and 6-prenyl-naringenin, isoxanthohumol, genistein, acacetin, chrysin and polyphenol – resveratrol; chalcones xanthohumol, isobavachalcone and licochalcone A and statins pravastatin, lovastatin, mevastatin and simvastatin [38, 55]. Importantly, the inhibitory effect on the channels is additive or synergistic when the flavonoids 8-prenylnaringenin, 6-prenylnaringenin, acacetin and chrysin and the chalcones xanthohumol and isobavachalcone are co-applied with the statins mevastatin and simvastatin [38, 55]. To the group of small-molecule organic compounds also belong newly designed thiophene-based inhibitors, recently designed by Gubic and co-workers (**Table 2**) [56].

6. More research required

As it was shown above, an application of the 'patch-clamp' technique, mainly in the whole-cell recording mode, enabled researchers to discover many various inhibitors of Kv1.3 channels, in both normal and cancer cells. An inhibition of the channels may be beneficial in treatment of a wide spectrum of diseases. However, in order to apply various inhibitors of Kv1.3 channels in medicinal practice, more research studies combining the 'patch-clamp' technique and non-electrophysiological approaches are required.

Studies on new compounds have to be performed. One of them is genistein, a plant-derived isoflavone known as a potent inhibitor of the protein tyrosine kinase (PTK). Genistein is a biologically active compound, which exerts pleiotropic effects including anti-proliferative and pro-apoptotic activities on Kv1.3 channel-expressing cancer cells [55]. It was shown that genistein is an inhibitor of Kv1.3 channels in normal human T lymphocytes [45]. Recent preliminary studies provide evidence that genistein also inhibits Kv1.3 channels expressed in human T cell line Jurkat [Teisseyre, preliminary results]. The inhibitory effect of genistein on the channels occurred in a concentration-dependent manner. It was accompanied by a significant slowing of the channel activation rate without significantly affecting the inactivation. The inhibitory effect of genistein on the channels was reversible for all applied concentrations. Obtained results may suggest that genistein molecules directly bind to the channel protein and stabilise the channel in its closed state, leading to a delay of its opening upon an application of an activating (opening) stimulus. The delayed opening of the channels may be responsible for an apparent reduction of the whole-cell currents upon an application of genistein. Genistein molecules probably dissociate from the channel protein upon its opening and re-associate with the channel after its closure [Teisseyre, preliminary results]. The inhibitory effect of genistein on Kv1.3 channels in Jurkat T cells is significantly augmented upon a co-application with mevastatin [Teisseyre, preliminary results]. In order to study the mechanism of the channel inhibition by genistein, more studies will have to be done, including the docking analysis. A combination of electrophysiological and non-electrophysiological techniques is necessary to study a putative relationship between the channel inhibition and anti-cancer activities of genistein.

Other naturally occurring compounds to be tested are flavonoids morin, hesperetin, quercetin and kaempferol; chalcones xanthohumol and isobavachalcone and a polyphenol resveratrol. All the flavonoids listed above are inhibitors of voltage-gated potassium channels Kv11.1 expressed in *Xenopus* oocytes [57, 58]. The potency of the Kv11.1 channel inhibition is the highest for morin and the lowest for quercetin and kaempferol [57, 58]. Moreover, hesperetin is an inhibitor of Kv1.5 channels expressed in

HEK293 cells [57]. Quercetin is also an activator of Kv1.5 channels expressed in *Xenopus* oocytes and in rats '*in vivo*' and an inhibitor of these channels in HEK293 cells [57]. Xanthohumol and isobavachalcone are already known as inhibitors of Kv1.3 channels in cancer Jurkat T cells [30, 38, 46, 55]. More studies combining the 'patch-clamp' and non-electrophysiological techniques are necessary to elucidate putative relationship between inhibition of the channels and anti-cancer activity of these compounds [55]. Resveratrol is known as an inhibitor of K1.3 channels in normal human T lymphocytes [30, 38]. A putative inhibitory effect of this compound on the channels expressed in cancer cells remains to be studied in detail, as well as the relationship between the channel inhibition and anti-cancer activity. Among statins, the best candidates for further studies are cerivastatin, fluvastatin and atorvastatin. All these compounds share anti-cancer activities against Kv1.3 channel-expressing cancer cells [55]. The selected flavonoids, chalcones and polyphenols will be applied alone and in combination with the statins [55].

7. Conclusions

The 'patch-clamp' technique is a useful electrophysiological technique that can be applied to record activity of ion channels on both the single-channel and the whole-cell level. An application of the later configuration is necessary in case of small cells, which cannot be examined applying the classical two-microelectrode voltage-clamp technique. One of types of voltage-gated potassium channels, which were discovered in small cells, such as human T lymphocytes, applying the whole-cell 'patch-clamp' technique, are Kv1.3 channels. These channels are widely expressed among tissues, both normal and cancer. The channels play an important role in regulation of the cell life and death. Inhibition of the channels may be beneficial in treatment of numerous diseases. Inhibitors of Kv1.3 channels may putatively find an application in a medicinal practice. In order to accomplish this goal, more combined studies with an application of the 'patch-clamp' and non-electrophysiological techniques will have to be performed.

Acknowledgements

The authors want to express the best thanks to Mrs. Anna Uryga for a successful research co-operation.

Conflict of interest

The authors have no conflict of interest.

Author details

Andrzej Teisseyre*, Kamila Środa-Pomianek and Anna Palko-Labuz
Department of Biophysics and Neurobiology, Medical University in Wrocław, Wrocław, Poland

*Address all correspondence to: andrzej.teisseyre@umw.edu.pl

References

[1] Hille B. Ionic Channels of Excitable Membranes. Sunderland, MA: Sinauer Associates Inc.; 2001

[2] Bernstein J. Untersuchungen zur Thermodynamik der bioelektrischen Ströme. Erster Theil Pflügers Arch. 1902;**92**:521-562

[3] Bernstein J. Elektrobiologie. Braunschweig: Vieweg; 1912

[4] Herrmann L. Lehrbuch der Physiologie. Berlin: August Hirschwald; 1905

[5] Herrmann L. Beiträge zur Physiologie und Physik des Nerven. Pflügers Archiv. 1905;**109**:95-144

[6] Young J. Structure of nerve fibres and synapses in some invertebrates. Cold Spring Harbor Symposia on Quantitative Biology. 1936;**4**:1-6

[7] Hodgkin A. Evidence for electrical transmission in nerve part I and II. Journal of Physiology (London). 1937;**90**(183-210):211-232

[8] Cole K, Curtis H. Electrical impedance of the squid giant axon during activity. The Journal of General Physiology. 1939;**22**:649-670

[9] Hodgkin A, Huxley A. Action potentials recorded from inside a nerve fibre. Nature (London). 1939;**144**:710-711

[10] Curtis H, Cole C. Membrane action potentials from the squid giant axon. Journal of Cellular and Comparative Physiology. 1940;**15**:147-152

[11] Curtis H, Cole K. Membrane resting and action potentials from the squid giant axon. Journal of Cellular and Comparative Physiology. 1942;**19**:135-144

[12] Hodgkin A, Katz B. The effect of sodium ions on the electrical activity of the giant axon of the squid. Journal of Physiology (London). 1949;**108**:37-77

[13] Marmont C. Studies on the axon membrane. I. A new method. Journal of Cellular and Comparative Physiology. 1949;**34**:351-382

[14] Cole K. Dynamic electrical characteristics of the squid axon membrane. Archives des Sciences Physiologiques. 1949;**3**:253-258

[15] Hodgkin A, Huxley A, Katz B. Measurements of current-voltage relations in the membrane of the giant axon of *Loligo*. Journal of Physiology (London). 1952;**116**:424-448

[16] Hodgkin A, Huxley A. Currents carried by sodium and potassium ions through the membrane of the giant axon of *Loligo*. Journal of Physiology (London). 1952;**116**:449-472

[17] Hodgkin A, Huxley A. The components of membrane conductance in the giant axon of *Loligo*. Journal of Physiology (London). 1952;**116**:473-496

[18] Hodgkin A, Huxley A. The dual effect of membrane potential on sodium conductance in the giant axon of *Loligo*. Journal of Physiology (London). 1952;**116**:497-506

[19] Hodgkin A, Huxley A. A quantitative description of membrane current and its application to conduction and excitation in nerve. Journal of Physiology (London). 1952;**117**:500-544

[20] Neher E, Sakmann B. Single-channel currents recorded from membrane of denervated frog muscle fibres. Nature (London). 1976;**260**:799-801

[21] Neher E, Sakmann B, Steinbach J. The extracellular patch-clamp: A method for resolving currents through individual open channels in biological membranes. Pflügers Archiv. 1978;**375**:219-228

[22] Sigworth F, Neher E. Single Na+ currents observed in cultured rat muscle fibres. Nature (London). 1980;**287**:447-449

[23] Langmuir I. Overturning and anchoring of monolayers. Science. 1938;**87**:493-500

[24] Hamill O, Marty A, Neher E, Sakmann B, Sigworth F. Improved patch-clamp techniques for high-resolution current recording from cells and cell-free membrane patches. Pflügers Archiv. 1981;**391**:85-100

[25] Sakmann B, Neher E. Single channel recording. New York: Plenum; 1983

[26] Sigworth F. The variance of sodium current fluctuations at the node of Ranvier. Journal of Physiology (London). 1980;**307**:97-129

[27] Matteson D, Deutsch C. K+ channels in T lymphocytes: A patch-clamp study using monoclonal antibody adhesion. Nature. 1984;**307**:468-471

[28] Cahalan M, Chandy K, DeCoursey T, Gupta S. A voltage-gated potassium channel in human T lymphocytes. Journal of Physiology. 1985;**358**:197-237

[29] Gutman G, Chandy KG, Grissmer S, Lazdunski M, McKinnon D, Pardo L, et al. International Union of Pharmacology. LIII. Nomenclature and molecular relationships of voltage-gated potassium channels. Pharmacological Reviews. 2005;**67**:473-508

[30] Teisseyre A, Palko-Labuz A, Środa-Pomianek K, Michalak K. Voltage-gated potassium channel Kv1.3 as a target in a therapy of cancer. Frontiers in Oncology. 2019;**9**:1-16. DOI: 10.3389/fonc.2019.00933

[31] Szabo I, Bock J, Jekle A, Soddemann M, Adams C, Lang F, et al. A novel potassium channel in lymphocyte mitochondria. Journal of Biological Chemistry. 2005;**280**(**13**):12790-12798

[32] Gulbins E, Sassi N, Grassme H, Zoratti M, Szabo I. Role of Kv1.3 mitochondrial potassium channel in apoptotic signalling in lymphocytes. Biochimica et Biophysica Acta (Bioenergetics). 2010;**1797**:1251-1259

[33] Feske S, Wulff H, Skolnik E. Ion channels in innate and adaptive immunity. Annual Reviews of Immunology. 2015;**33**:291-353

[34] Perez-Verdaguer M, Capera J, Serrano-Novillo C, Estadella I, Sastre D, Felipe A. The voltage-gated potassium channel Kv1.3 is a promising multitherapeutic target against human pathologies. Expert Opinion on Therapeutic Targets. 2016;**20**(5):577-591

[35] Chandy KG, Norton R. Peptide blockers of Kv1.3 channels in T cells as therapeutics for autoimmune diseases. Current Opinion in Chemical Biology. 2017;**38**:97-107

[36] Serrano-Albarras A, Estadella I, Cirera-Rocosa S, Navarro-Perez FA. Kv1.3: A multifunctional channel with many pathological implications. Expert Opinion on Therapeutic Targets. 2018;**22**(2):101-105

[37] Bachmann M, Li W, Edwards M, Ahmad S, Patel S, Szabo I, et al. Voltage-gated potassium channels as regulators of cell death. Frontiers in cell and Developmental Biology. 2020;**8**:article 611853

[38] Teisseyre A, Środa-Pomianek K, Palko-Labuz A, Chmielarz M. Voltage gated potassium channels Kv1.3 in health and disease. In: Chapter of a Book: Physiology, Annual Volume 2024. London, UK: IntechOpen; 2023. DOI: 10.5772/intechopen.113769

[39] Jusztus V, Medyouni G, Bagosi A, Lampe R, Panyi G, Matolay O, et al. Activity of potassium channels in $CD8^+$ T lymphocytes: Diagnostic and prognostic biomarker in ovarian cancer. International Journal of Molecular Sciences. 2024;**25**:1949. DOI: 10.3390/ijms25041949

[40] Nguyen H, Grössinger E, Horiuchi M, Davis K, Jin L, Maezawa I, et al. Differential Kv1.3, K(Ca)3.1 and Kir 2.1 expression in "classically" and "alternatively" activated microglia. Glia. 2017;**65**:106-121

[41] Felipe A, Vincente R, Villalonga N, Roura-Ferrer M, Martinez-Marmol R, Sole L, et al. Potassium channels: New targets in cancer therapy. Cancer Detection and Prevention. 2006;**30**:375-385

[42] Bielanska J, Hernandez-Losa P-VM, Moline T, Somoza R, Ramon y Cajal S, Condom E, Ferreres J, Felipe A. Voltage-dependent potassium channels Kv1.3 and Kv1.5 in human cancer. Current Cancer Drug Targets. 2009;**9**:904-914

[43] Felipe A, Bielanska J, Comes N, Vallejo A, Roig S, Ramon y Cajal S, et al. Targeting the voltage-gated K+ channels Kv1.3 and Kv1.5 as tumor biomarkers for cancer detection and prevention. Current Medicinal Chemistry. 2012;**19**:661-674

[44] Comes N, Bielanska J, Vallejo-Garcia A, Serrano-Albarras A, Marruecos L, Gomez C, et al. The voltage-dependent K+ channels Kv1.3 and Kv1.5 in human cancer. Frontiers in Physiology. 2013;**4**(Article 283):1-12

[45] Teisseyre A, Michalak K. Genistein inhibits the activity of Kv1.3 potassium channels in human T lymphocytes. Journal of Membrane Biology. **2005**;**205**:71-79

[46] Teisseyre A, Uryga A, Środa-Pomianek K, Palko-Łabuz A. The influence of isobavachalcone and selected non-prenylated chalcones on the activity of Kv1.3 channel in the cancer cell line Jurkat T. Cellular & Molecular Biology Letters. manuscript submitted

[47] Wang X, Li G, Guo J, Zhang Z, Zhang S, Zhu Y, et al. Kv1.3 channel as a key therapeutic target for neuroinflammatory diseases: State of the art and beyond. Frontiers in Neuroscience. 2020;**13**:article 1393. DOI: 10.3389/fnins2019.01393

[48] Kazama I. Physiological significance of delayed rectifier K+ channels (Kv1.3) expressed in T lymphocytes and their pathological significance in chronic kidney disease. The Journal of Physiological Sciences. 2015;**65**:25-35. DOI: 10.1007/s12576-014-0331-x

[49] Kazama I, Tamada T, Tachi M. Usefulness of targeting lymphocyte Kv1.3 channels in treatment of respiratory diseases. Inflammation Research. 2015;**64**:753-765. DOI: 10.1007/s00011-015-0855-4

[50] Kazama I, Maruyama Y, Murata Y, Sano M. Voltage-dependent biphasic effects of chloroquine on delayed

rectifier K+ channel currents in murine thymocytes. The Journal of Physiological Sciences. 2012; **62:** 267-274

[51] Checchetto V, Prosdocini E, Leanza L. Mitochondrial Kv1.3: A new target in cancer biology? Cellular Physiology and Biochemistry. 2019;**53**:52-62

[52] Prosdocimi E, Checchetto V, Leanza L. Targeting the mitochondrial potassium channel Kv1.3 to kill cancer cells: Drugs, strategies and new perspectives. SLAS Discovery. 2019;**24**:882-892. DOI: 10.1177/2472555219864894

[53] Kadow S, Schumacher F, Kramer M, Hessler G, Scholtysik S, Oubari S, et al. Mitochondrial Kv1.3 channels as target for treatment of multiple myeloma. Cancers. 2022;**14**:1-16. DOI: 10.3390/cancers14081955

[54] Li W, Wilson G, Bachmann M, Wang J, Mattarei A, Paradisi C, et al. Inhibition of mitochondrial potassium channel in combination with gemcitibane and abraxane drastically reduces pancreatic ductal adenocarcinoma in an immunocompetent orthotopic murine model. Cancers. 2022;**14**:1-23. DOI: 10.3390/cancers14112618

[55] Teisseyre A, Środa-Pomianek K, Palko-Labuz A. The influence of naturally occurring flavonoids, chalcones and statins on the activity of voltage-gated potassium channels and viability of Kv1.3 channel-expressing cancer cells. Journal of Molecular Structure. 2024;**1307**:1-11. DOI: 10.106/j.molstruc.2024.137967

[56] Gubic S, Hendricks L, Shi X, Toplak Z, Mozina S, van Theemsche K, et al. Design a new potent and selective thiophene-based Kv1.3 inhibitors and their potential for anticancer activity. Cancers. 2022;**14**:2595. DOI: 10.3390/cancers14112595

[57] Richter-Laskowska M, Trybek P, Delfino V, Wawrzkiewicz-Jałowiecka A. Flavonoids as modulators of potassium channels. International Journal of Molecular Sciences. 2023;**24**:311. DOI: 10.3390/ijms24021311

[58] Zitron E, Scholz E, Owen R, Lück S, Kiesecker C, Thomas D, et al. QTc prolongation by grapefruit juice and its potential pharmacological basis: HERG channel blockade by flavonoids. Circulation. 2005;**111**:835-838